T0140207

Lecture Notes in Information Systems and Organisation

Volume 22

More information about this series at http://www.springer.com/series/11237

Jerzy Gołuchowski · Małgorzata Pańkowska
Henry Linger · Chris Barry
Michael Lang · Christoph Schneider
Editors

Complexity in Information Systems Development

Proceedings of the 25th International
Conference on Information Systems
Development

 Springer

Editors
Jerzy Gołuchowski
Department of Knowledge Engineering
University of Economics in Katowice
Katowice
Poland

Małgorzata Pańkowska
Department of Informatics
University of Economics in Katowice
Katowice
Poland

Henry Linger
Caulfield School Information Technology
Monash University
Caulfield East, VIC
Australia

Chris Barry
Cairnes School of Business & Economics
National University of Ireland
Galway
Ireland

Michael Lang
Cairnes School of Business & Economics
National University of Ireland
Galway
Ireland

Christoph Schneider
Department of Information Systems
City University of Hong Kong
Kowloon
Hong Kong

ISSN 2195-4968   ISSN 2195-4976   (electronic)
Lecture Notes in Information Systems and Organisation
ISBN 978-3-319-52592-1   ISBN 978-3-319-52593-8   (eBook)
DOI 10.1007/978-3-319-52593-8

Library of Congress Control Number: 2017933067

Printed on acid-free paper

This Springer imprint is published by Springer Nature
The registered company is Springer International Publishing AG
The registered company address is: Gewerbestrasse 11, 6330 Cham, Switzerland

Preface

The **International Conference on Information Systems Development** (ISD) is an academic conference where researchers and practitioners share their knowledge and expertise in the field of information systems development. As an Affiliated Conference of the Association for Information Systems (AIS), the ISD conference complements the international network of general IS conferences (ICIS, ECIS, AMCIS, PACIS, ACIS). The ISD conference continues the tradition started with the first Polish-Scandinavian Seminar on Current Trends in Information Systems Development Methodologies, held in Gdansk, Poland in 1988. This seminar has evolved into the International Conference on Information Systems Development.

During its history, the conference has focused on different aspects, ranging from methodological, infrastructural, or educational challenges in the ISD field to bridging the gaps between industry, academia, and society. The development of information systems has paralleled technological developments and the deployment of those technologies in all areas of society, including government, industry, community, and in the home. ISD has always promoted a close interaction between theory and practice that has set an agenda focused on the integration of people, processes, and technology within a specific context.

This publication is a selection of papers from ISD2016, the 25th ISD conference hosted by the Faculty of Informatics and Communication, University of Economics in Katowice, and held in Katowice, Poland on August 24–26, 2016. All accepted papers have been published in the AIS eLibrary at http://aisel.aisnet.org/isd2014/proceedings2016. This volume contains extended versions of the best papers, as selected by the ISD2016 Proceedings Editors.

The theme of the conference was *Complexity in Information Systems Development*. Complexity in information systems development includes considerations of context, creativity, and cognition in the information systems development field which enable information systems to be examined as they are actually developed, implemented, and used. Context-aware computing refers to the ability of computing devices to detect and sense, interpret, and respond to, aspects of a user's local environment and the computing devices themselves. Thus, considerations of context are especially important for mobile applications development. Cognition and creativity are also present in an organization's development efforts, in the identification of new markets and technologies, and in their information systems projects.

Context, creativity and cognition have different interpretations in social and technical sciences, which encourages inter-, cross-, multi- and intradisciplinary research. As such, complexity in information system development is manifested by projects and research works focusing on technological issues as well as on social, organizational and economic issues.

ISD2016 focused on these and associated topics in order to promote research into theoretical and methodological issues and ways in which complexity influence the field of Information Systems Development. We believe that the papers assembled in this publication are an important contribution in this regard.

Katowice, Poland	Jerzy Gołuchowski
Katowice, Poland	Małgorzata Pańkowska
Caulfield East, Australia	Henry Linger
Galway, Ireland	Chris Barry
Galway, Ireland	Michael Lang
Kowloon, Hong Kong	Christoph Schneider

Conference Organization

Conference Chair

Jerzy Gołuchowski

Programme Chairs

Małgorzata Pańkowska

Local Organising Committee

Edyta Abramek
Barbara Filipczyk
Anna Kempa
Joanna Palonka
Anna Sołtysik-Piorunkiewicz
Artur Strzelecki
Wiesław Wolny
Mariusz Żytniewski
Artur Machura
Mariia Rizun
Dominika Zygmańska

International Steering Committee

Chris Barry
Michael Lang

Henry Linger
Christoph Schneider

Track Chairs

Information Systems Methodologies and Modelling
Igor Hawryszkiewicz
Paulo Rupino da Cunha
Kieran Conboy

Managing IS Development

Emilio Insfran
William Wei
Asif Qumer Gill

ISD Education

Marite Kirikova
John Traxler
Mark Freeman

Context-awareness in ISD

Samia Oussena
Dumitru Dan Burdescu
Mieczyslaw Owoc

Cognitive Science

Jaroslav Pokorny
Joan Lu
Nina Rizun

Creativity Support Systems

Celina Olszak
Petros Kostagiolas
Maciej Nowak

Greening by IT Versus Green IS

Claus-Peter Rückemann
Fabio De Felice
Dr. G.P. Sahu

General Concepts (including Legal and Ethical Aspects of ISD, Project Management, and Other Topics)

Tihomir Orehovački
Chris Barry
Michael Lang

Reviewers

Witold Abramowicz
Silvia Abrahao
Per Backlund
Chris Barry
Peter Bellström
Paul Beynon-Davies
Alena Buchalcevová
Dumitru Dan Burdescu
Frada Burstein
Dave Bustard
Andrzej Bytniewski
Jenny Coady
Kieran Conboy
Witold Chmielarz
Alfredo Cuzzocrea
Liam Doyle
Helena Dudycz
Dariusz Dziuba
María José Escalona

Odd Fredriksson
Mark Freeman
Stéphane Galland
Shang Gao
Javier Gonzalez
Mariusz Grabowski
Igor Hawryszkiewicz
Emilio Insfran
Dorota Jelonek
Karlheinz Kautz
Leszek Kieltyka
Rónán Kennedy
Marite Kirikova
Jerzy Kisielnicki
Andrzej Kobylinski
Jerzy Korczak
Petros Kostagiolas
Michael Lang
Jay Liebowitz

Contents

A Conceptual Investigation of Maintenance Deferral and Implementation: Foundation for a Maintenance Lifecycle Model

Christopher Savage, Karlheinz Kautz and Rodney J. Clarke

1 Introduction

The "dependence on critical infrastructures is increasing worldwide" [1, p. 112] and "both the impact of software on life, and our dependence on software is rapidly increasing" [2, p. 531]. Companies requiring software capability that do not want to develop the capability in-house can choose to commission or outsource a unique build, or purchase the capability [3]. By purchasing from a vendor, the organization "benefits [from] generic best practices and advanced functionality supported by vendors' research capabilities" [4, p. 219]. Over time, purchasing this capability has become increasingly attractive [5] and "once an organization has adopted packaged software, upgrades to newer versions are inevitable" [3, p. 153]. The focus of this research is on vendor-supplied standard packaged software, abbreviated hereafter to vendor software, which is considered to be generic software, pre-created by a third-party organization for the purpose of sale or licensing. Vendor software is treated as including 3rd-party, commercial-off-the-shelf (COTS) software [5], product software [2] or packaged software [3].

A prior version of this paper has been published in the ISD2016 Proceedings (http://aisel.aisnet.org/isd2014/proceedings2016).

C. Savage · R.J. Clarke
University of Wollongong, Wollongong, Australia
e-mail: cns993@uowmail.edu.au

R.J. Clarke
e-mail: rclarke@uow.edu.au

K. Kautz (✉)
Royal Melbourne Institute of Technology (RMIT) University, Melbourne, Australia
e-mail: karlheinz.kautz@rmit.edu.au

© Springer International Publishing Switzerland 2017
J. Gołuchowski et al. (eds.), *Complexity in Information Systems Development*,
Lecture Notes in Information Systems and Organisation 22,
DOI 10.1007/978-3-319-52593-8_1

IEEE defines software maintenance as "the process of modifying a software system or component after delivery to correct faults, improve performance or other attributes, or adapt to a changed environment" [6, p. 46] while Swanson refers to maintenance as "all modifications made to an existing application system, including enhancements and extensions" [7, p. 311]. For this research we adopt a definition of maintenance both as a process and as the outcome of that process. Vendors will periodically deliver maintenance to the purchasing organization in the form of patches or upgrades ready to be applied to installed systems. The vendor develops and releases the maintenance, but each purchasing organization may have to expend significant effort to incorporate the maintenance into the production environment which may lead to the "typical option of 'doing nothing'" [8, p. 451], "its usual preference to 'ride [the current version] out as long as possible'" [9, p. 562] in which "neglect is the inertially easy path" [1, p. 112]. This conscious or unconscious decision to postpone or delay implementation of the vendor-supplied maintenance into the operational environment is considered deferral within this paper. The study takes the purchaser's viewpoint and explores the current state of literature within the topic of maintenance deferral of vendor software. It identifies the reasons for deferral and performance of vendor-supplied maintenance by the purchasing organization. These reasons have previously not been studied from that perspective. The identified reasons build the groundwork and foundation for a Maintenance Lifecycle model that provides a starting point to research vendor-supplied maintenance from the customer's point of view.

The paper is structured as follows: some background of software maintenance and maintenance deferral is provided in the next section, followed by the description of the applied literature review method. The key themes and concepts are identified from the application of this review. A Maintenance Lifecycle Model is deduced from the literature review. A discussion and conclusion section completes the paper with a summary highlighting a key gap that provides an opportunity for further research while stating the limitations of the current research.

2 Background

Following the purchase of vendor software, the software needs support in order to maximize its operational life because "Systems are nevertheless subject to structural deterioration and obsolescence with age" [10, p. 278]. By installing vendor software, purchasers will have to "be prepared for managing the impacts of [maintenance]" [3, p. 167]. The cost of purchasing vendor maintenance is not a concern as many vendors employ license agreements whereby maintenance is made available without additional charge [11]. The comprehensive view on maintenance which we embrace based on Swanson [7] results in the inclusion of major and minor upgrades, patches and maintenance within this study. The maintenance period is commonly referred to as being the longest phase of the software lifecycle [5, 12]. The maintenance period for vendor software begins during commissioning, as vendor-supplied maintenance is incorporated into the commissioning in order to prevent the client starting operations with an "out of date" system [13].

Within this study, deferral is treated as a conscious or unconscious decision of the purchasing organization that postpones or delays a course of action. Implicit within this definition of deferral is that the postponed action will have to be performed at some future time. Referring to road maintenance, Harvey [14, p. 34] captures the essence of maintenance deferral in any realm as "deferring maintenance can be seen as a form of borrowing. Funds are saved in the short-term at the expense of higher outlays in the future."

Deferral becomes a critical issue for the purchaser of vendor software when the vendor declares an "end of life" (EOL) date, indicating that further maintenance ceases for this version [15]. This forces the business to accept a new risk of using a component of unsupported IT infrastructure or software, or perform maintenance to move onto a supported version of software [9]. In reviewing papers relating to the deferral of vendor-supplied maintenance, this paper will investigate how EOL can become a problem for purchasers of vendor software due to the purchaser repeatedly deferring the adoption of newer versions of the software. The backlog of software maintenance activities for software packages creates a poorly-understood risk for organizations [13] and warrants further research.

3   Review Method

Our conceptual investigation is based on a systematic literature review. The execution of the review and the structure of reporting its results follow a concept-centric literature review approach [16] to ensure that repeatable data gathering and logical analysis support the discussions presented and conclusions drawn. Kitchenham and her associates [17, 18] provided guidance for a systematic review. The review progresses through a sequence of filtering results as presented by [19]. The addition of a preliminary, informal step expands the number of initially considered papers, thereby reducing the risk of accidently eliminating papers during the initial search [18].

In preparation for the systematic review, an unstructured review of publically available literature through a State Library was conducted using the terms "maintenance deferral", "project prioritization" and "project prioritisation". Iterative snowball addition of key words and concepts from the resulting papers created 10 different search terms related to the core concepts listed above. To maximize the scope of literature, the initial search was not limited to topic-specific databases, popular publications or peer-reviewed papers. This is consistent with the advice that a wide net should be cast in order to consider all published articles in a field [16].

The Web of Science™ database was selected for this review due to the wide cross-discipline nature of the index including the Association for Information Systems' Senior Scholars "basket of 8" journals and all but two journals from The Financial Times 45 top journal list. For this review, the titles of about 14900 papers were evaluated. Through the different screening processes a total of 40 papers were included into the review. The results of this analysis are presented in the next section.

4   Results

Despite the broad search terms, every item that passed the critical review referred to the maintenance deferral problem from a vendor's perspective; no papers addressed the problem directly from the purchaser's perspective. The papers were published over a period of nearly 40 years with the first one appearing in 1977. This demonstrates that the topic of maintenance deferral is not new. The geographical distribution of papers indicates that the deferral problem discussed here provides a "Western" view of the issue and may not be globally generalizable. The literature review filtering criteria of English-language papers will have influenced this distribution. Five concepts and themes emerged from the literature; they are: (1) maintenance of vendor software is a problem, (2) there is too little research on the topic, (3) reasons for maintenance deferral do exist, (4) deferral has consequences (5) there are reasons for maintenance implementation.

4.1   Maintenance of Vendor Software Is a Problem

The literature acknowledges that adoption of vendor software causes a maintenance problem for the purchasing organization [1, 8, 9, 13, 20]. No paper expressed a dissenting opinion that vendor software is free of maintenance impacts. Within this acknowledgement, several reasons which we will discuss in the following subsections were identified that led to organizational caution when assessing vendor-supplied maintenance before implementing it into production environments.

4.2   There Is Too Little Research on the Topic

The literature includes numerous calls for further investigation into the maintenance of vendor-supplied systems as well as for the maintenance deferral phenomenon and highlights the increasing issue of maintenance backlog in IT systems and infrastructure.

Already 1983 Lientz [21, p. 277] requested "much more research is needed in maintenance". In 1995 Swanson stated "I wouldn't do the same [1979 Dimensions of Software Maintenance] study [today]. ... I would try to focus on the maintenance of commercial software packages ... Or, I would address maintenance from the user perspective, which has been largely ignored." [7, p. 307]. In the same vein several authors [3, 9, 20, 22] lament the neglect of investigations into vendor software maintenance. Khoo et al. [22, p. 334] implicitly call for more research stating

"Although our research provides an initial investigation into the phenomenon of support upgrades, the empirical support for our findings were limited to a single upgrade case."

Hybertson et al. [23, p. 215] similarly state that "COTS use is increasing, and maintenance issues of COTS-intensive systems need to be articulated and addressed." They are supported by Reifer et al. [15, pp. 95, 96] who say "Currently, few COTS software lifecycle models address [Component-Based System] maintenance processes" and demand "To make better decisions relative to [Component-Based Systems], we need empirical knowledge. To gain this knowledge, we must understand more fully the lifecycle processes people use when harnessing COTS packages." The absence of academic framework(s) or in-depth research addressing the organizational behavior during the period between the vendor publishing maintenance to the purchaser and the tipping point that triggers the maintenance to be applied to the purchaser's system(s) is also stated in [3].

Literature relating to the initial investment decision and deriving the full expected benefits from a past investment decision were prevalent, an observation supported by [22]; however software maintenance is either mostly ignored both by research and practice [24] or is simply not attractive, considered "less glorious" [25] and suffering a negative image with developers and managers involved in the process [26–28].

Finally, only few papers did employ theoretical models to describe some aspects of the vendor-supplied maintenance deferral issues. Communicative framing theory is used to show how consistent messages and actions prepared and supported users through the application of a major IS maintenance activity through the use of a galvanizing negatively-framed message [22]. An inductive research strategy and comparative analysis is presented in [9] to construct a theoretical model about the interaction of factors influencing upgrade decisions by adopting a critical realism approach to explain motives, contingencies and dependencies impacting the decisions. Finally, Khoo et al. [3] extend Swanson and Beath's [25] Relational Foundation model to incorporate the vendor relationships in an explanation of the impacts that vendor software upgrades have on business and IS stakeholders. Hanna and Martin [29] discuss a model that incorporates vendor-supplied maintenance into a larger Repair Level Analysis, but complain that IS researchers and practitioners have so far failed to embrace such modeling within pure IT systems. In summary, there is too little research into the maintenance of vendor-supplied systems.

4.3 Reasons for Maintenance Deferral Do Exist

From the literature, a common theme of reasons for maintenance deferral can be deduced. In almost all cases, analysis of these reasons often expressed as risks suggests that their consequences can be avoided through the deferral of vendor-supplied maintenance, or exercising the 'doing nothing' option. Table 1 presents the reasons for deferral of maintenance expressed across literature assessed for this conceptual study.

Table 1 Reasons for maintenance deferral

Loss of customizations, configurations, interfaces	Complicated and expensive test environments and infrastructure
Huge costs	Disrupting to the organization and productivity
Chain reaction of cascading maintenance	Unforeseen impacts, impossible to complete tests
	Dependence on vendor claims of suitability
Training efforts and steep user learning curve	Dependence on vendor documentation
	Conflict with the vendor
Poor quality, conflict for existing and new	Resistance and user revolt
IS resources	Additional work for expert users to train others
Effort to analyse, test, or implement	Requiring a re-certification for a certified system
Inconvenient rate and time of arrival	
Disturbing the existing IS equilibrium	
Difficult or complex	

The risk of losing customizations, configurations, or interfaces was most recognizable in the literature. It extends beyond the technology-based concerns into the realm of the user as "Users also create idiosyncratic adaptations and workarounds to overcome limitations in any customized software" [22, p. 329] that could be impacted through the application of maintenance.

Almost as prevalent was the risk that vendor-supplied maintenance would have a huge cost associated with it [13]. As purchasing organizations implement vendor-supplied systems to gain a commercial advantage [20] any planned or unplanned expense in monetary or effort-based terms may detract from this profit-making goal. In some of the few direct references to deferral, cuts and limits in maintenance budgets are a common occurrence and the flow-on deferral of maintenance is a direct result [15, 23]. A more general economic downturn may also lead to maintenance being seen as too costly [30].

The risk that maintenance to one system will cause a chain reaction of integration updates and backward-compatibility issues has been very common in the literature. Both minor inconveniences such as missing device drivers following operating system maintenance requiring replacement of printers, faxes and scanners [3] and a case of thirteen linked vendor-supplied systems requiring upgrade [31] related to this risk. The risk of cascading maintenance applies also to internal maintenance requirements of a vendor-supplied system. A mandatory maintenance action on one module may cause issues requiring further maintenance of a separate module [13].

A further reason that appears regularly in the literature is related to training effort and user's learning curves. Khoo et al. [22, p. 332] state "Because SAP upgrades usually involved downtime and training, business users normally preferred to defer an upgrade as long as possible." This is supported by [9, 13].

The risk that a vendor-supplied maintenance release will be of poor quality and introduce bugs and conflicts between existing and new IS resources appears already in work of the early 1980s [32] and is frequently confirmed [7, 22].

The risk that a vendor-supplied maintenance release will consume a tremendous amount of effort to analyze, test, or implement is significant and justified. The need for testing is not eliminated through the implementation of vendor software [33] and the literature reports testing and implementation efforts between six months and a year for some upgrades [8, 22].

The unpredictable behavior of vendors, where "it is difficult to determine when the software will be released" [2, p. 533], or simply the "burdensome ... rate of change" for vendor software [5, p. 362] leads to risks concerning the inconvenient rate and time of arrival of maintenance releases.

A number of studies included the risk that maintenance will also disturb the existing equilibrium of information systems in the organization [1, 7, 20, 31]. A risk that a maintenance project can be as difficult and complex as the original installation is also existing [1, 3, 5, 20, 23].

Also related to unpredictable behavior of vendors is a risk of the unforeseen, related to the impossibility of fully testing a maintenance release and its un-assessable impacts and side-effects [2]. The risk of the unknown is implicitly common to all risks listed here, but there is a specific risk of the unforeseen, that even when everything is assessed and mitigated, something might go wrong with a concrete example of this unforeseen risk being pointed to as "Unexpected problems with file sharing in Access" in [7, p. 164].

An example of the risk of maintenance disrupting the organization and its productivity was the failure of a new feature in upgraded software causing "a mess for about three weeks" [3, p. 161]. A second example of organizational disruption [22] saw a company undergoing slowdown in performance and system lockouts subsequent to a three-day outage to implement the upgrade and in another case "files missing" [3, p. 162] as a side-effect from an upgrade.

Some papers lamented the complications and costs of maintaining environments for testing [5, 23, 32] as a specific risk with one recording three separate environments, development, test, and production, in an organization in order to manage and maintain their vendor-supplied system [13].

Because organizations need the vendor for the maintenance of vendor software, they must rely on the vendor claims concerning the suitability of the maintenance release; a risk pointed out by [5, 12, 31, 34, 35]. A similar requirement of dependence and a related risk is specifically valid for the documentation that accompanies a release as such documentation "might be incorrect on incomplete" [12, p. 13].

Inevitably, applying maintenance to an operational system may cause conflict with the vendor as illustrated by [7]: "During the ... testing phase, [Information Systems] staff identified many problems that they attributed to [the] software, but the vendor countered that the problems were related to client [organization] configuration decisions" [7, p. 165]. A risk of conflict with the vendor can be deduced from this and has been confirmed by [31, 34, 36]. A risk of maintenance leading to resistance and user revolt caused by changed software is also identified [3, 22]. Additional work for expert users in the form of training other employees is another reason for maintenance deferral [3]. Finally, a risk that upgraded software might require a re-certification for a certified system has been stated by [5].

4.4 Deferral Has Consequences

Deferral can be a logical, considered course of action when the risk of implementing the maintenance is calculated to be unacceptable [5]. An example of unacceptable risk is an incompatibility between the maintenance item and its environment, vendor disclosed [35] or otherwise identified or an identified threat to the stability of a system associated with a major release [9].

The consequences of maintenance deferral can otherwise be to avoid expense in the short term, however the legitimacy and suitability of this approach assume that no trigger event will occur. Should a trigger event occur and be ignored, possible consequences include economic damage to the company [38], higher expenditure and forced outages at a later time [30], or even demise of the purchasing organization itself [5].

Although IT maintenance can be deferred for one to two years, extended periods of deferral can lead to "the application portfolio risks getting dangerously out of date and a systemic risk" [39, p. 1]. The risks of systemic failure create a situation of positive-feedback where "the more different infrastructures that fail concurrently, the more difficult it becomes to restore service in any of them" [1, p. 112].

For organizations that have an understanding of the actual state of their systems, the act of maintenance deferral can be a considered action to save expense, and improve stability leading into a system retirement or replacement "as the end of any system's life is eventually foreseen, the maintenance effort itself may be moderated" [10, p. 279].

One possible consequence following repeated deferral is to completely separate from the vendor's support model and to "go it alone" through either maintaining the system in-house, or paying for bespoke support, possibly receiving a lower priority than up-to-date clients of the vendor [22]. However, the approach of deferring maintenance becomes precarious when vendor-supplied maintenance "that we require urgently" arrives, but has a dependency on a "backlog" of un-installed changes, which occurs because the vendor "seems to assume that you are up to date" [13, p. 100].

4.5 There Are Triggers for Maintenance

Mukherji et al. [40] put forward that investments in upgrades are best made when the gap between the new technology and the one currently in use reaches a critical threshold. A theme of identifiable trigger events, which cause this threshold to be reached immediately preceding the implementation of vendor-supplied

Table 2 Triggers and reasons for maintenance implementation

Need for increased business benefit	Standardization resulting from acquisition or merger
Avoid EOL date when vendor supports stops	Remain current with the marketplace
React on changed hardware requirements	Respond to a massive social change or innovation
Resolve an error relevant to the purchaser	
Satisfy company policy	Change to environment such as legislation
Standardization to remain compatible with external parties	React to release of vendor maintenance
	Eliminate or contain a security threat

maintenance, emerges from the literature. Table 2 summarizes the identified triggers and reasons for maintenance implementation.

Satisfying the need for increased business benefit is the most reoccurring theme within the literature concerning triggering the implementation of maintenance. This could be achieved through new functionality and features available within a newer release that fulfills user requests or requirements especially for improved performance. In this context, an aspiration of first-mover advantage also spurs maintenance [40]. Vendors declaring an EOL date for support of a particular version were an often referenced trigger for maintenance implementation [20]. The adoption of vendor software creates a lock-in situation where the purchasing organizations become dependent on the software vendor to provide them with software functionality and technical support [9]. This means that a vendor declaring an end to that support represents a significant risk to the purchasing organization. Lientz and Swanson [32] in their early studies on maintenance identified a non-software trigger event, the requirement to move from obsolescent hardware or to upgrade the hardware platform to mitigate hardware availability and support issues. This has been confirmed through the following years [31]. Several studies [10, 13, 20, 22, 23, 41] identify the resolution of an error relevant to the purchaser as a further trigger for maintenance. Policy within the organization aims to assist with determining the occurrence of a trigger event, however apparently contradictory policies with the same aim were identified in separate studies: one policy required a company remain within vendor-support version requirements [8], another one requested to upgrade every one and a half years [22]. A rationale for standardization as another trigger for maintenance is the need to remain compatible with external parties that interface an organization's information systems [37]. A need to standardize IS infrastructure resulting from a business acquisition or merger may also trigger maintenance [8].

A regular trigger for maintenance is the need to remain current with the marketplace [20, 21, 33, 40] as is change to the external environment such as legislation [21] to be dealt with legal-change-patches [20]. In addition, other environmental factors such as competitive pressure and general social and cultural factors were stated as triggers [21]. Major social change was also identified as triggering maintenance: e.g. the introduction of the Euro currency within the European Union [41]. Some papers alluded to the vendor maintenance release as a simple and sufficient trigger for a client organization's reaction to implement it [4, 13, 20, 33]. Finally, exploits or threats that increase the risk in a safety-critical, life-critical or secure system are possible triggers for maintenance [5, 38].

5 Deduction of a Maintenance Lifecycle

Beyond the identification of the reasons for maintenance deferral and implementation our literature review also uncovered concepts which allow us to put forward a maintenance lifecycle model. IEEE [6] puts forward a software lifecycle consisting of 8 phases, one of them being the operation and maintenance phase. It is defined as "the period of time in the software lifecycle during which a software product is employed in its operational environment, monitored for satisfactory performance, and modified as necessary to correct problems or to respond to changing requirements" [6, p. 52]. Through the synthesis of concepts spanning multiple critically reviewed papers, we deduce and propose a dedicated maintenance lifecycle from ideas not previously unified. This cycle begins with acquisition of the asset that creates a need to maintain the investment [1]; a trigger event causes maintenance to be required [5, 9]; the maintenance activity is planned [31]; the purchasing organization's IS and software users are prepared for the maintenance [22]; the maintenance is implemented; and the implications to the organization arising from the maintenance are stabilized [3]. Figure 1 uses the Software Lifecycle to demonstrate the placement of this maintenance lifecycle.

The proposed lifecycle is repetitive which refers to ongoing enhancements following the initial system commissioning and stabilization [21]. It is also reminiscent of Deming's Plan-Do-Act-Check cycle, in that it iterates through the states. The difference within the maintenance cycle is an explicit "wait" state before a

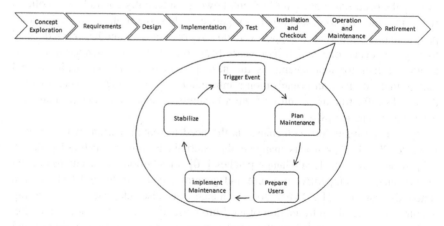

Fig. 1 A maintenance lifecycle model

trigger event, while the need for the next planning phase arises. In other words, there is not necessarily an automatic progression prior to the next trigger event.

6 Discussion

This study provides a summary of the reasons for maintenance deferral and implementation. It thereby advances three research problems stated by Gable et al. [34] in an early research framework for large packaged application software maintenance: (1) In exploring the rationale for deferral, the identified reasons refer to the drivers for a maintenance decision. (2) It takes up the question of to what extent can maintenance be avoided through packaged software solutions? Implicit in this question is the assumption that maintenance can be avoided, which is addressed by the deferral aspect within this study. (3) Lastly, a maintenance life-cycle model is deduced as a generic concept across all possible vendor-supplied systems and demonstrates that packaged software maintenance concepts are in fact generic and extensible beyond a particular vendor's product.

The study had to address several challenges. The first challenge was gaining a suitable view of the concept of software. For as long as IT, IS, and software investments have existed, there have been attempts to classify the artifacts in a way common to other investments. Through the adoption of a technical vendor view-point they can be divided into infrastructure, tools or applications [2]. An alternative view is to understand IT, IS and software as representing an asset [41]. This treatment of software as an asset supports references within the study to deferral behaviors within the engineering realm and its physical assets. The next hurdle was the very definition of the key term maintenance. The reviewed papers provided conflicting definitions of maintenance, patches and upgrades that confused attempts to synthesize a clear picture of the topic. This was reinforced through the diverse search terms required to capture the documents assessed. Therefore the IEEE [6] definition of maintenance as a process and Swanson's [7] view of maintenance as an outcome or product were adopted. Although vendor-supplied software maintenance can add new functionality, the customer judges the maintenance as required in order for the software asset to remain useful. The study is then based on an understanding that the maintenance phase begins following the purchase of vendor software, not its activation. This is because the first maintenance releases "will occur before the system's initial delivery" [33, p. 53].

Deferral has negative connotations as "deferred maintenance is an exercise that often detracts from the more fundamental task of attacking the problem itself. Because of the implication that deferral has been caused by neglect and not by conscious planning, administrators shy away from approaching the main job" [42, p. 43]. This study demonstrates that deferral has both legitimate and neglect-based causes. From the first origins of maintenance with the creation of tools and structures, items were used until they failed [43]. Through the industrial revolution, systems became more complex, but maintenance, apart from routine lubrication, remained largely something performed at the point of failure. During World War II the need for operational fighter aircrafts created a need for preventative maintenance, maintenance before failure, therefore creating a function to support availability. From the results of this study, the need for a trigger event before implementing maintenance strongly indicates that some IS owners are still behaving in a mode of operating-to-failure or, operating-to-obsolesce their software investments.

The need for different approaches to internal processes caused by the move from a traditional in-house development team to vendor software is poorly understood and provides a lens to understand the vendor-supplied maintenance deferral question [33]. It is possible that some purchasing organizations fail to make the change to utilizing packaged software based processes and therefore remain unaware of the pervasive ramifications both to people and processes that are triggered by implementing a vendor-supplied product and subsequent maintenance. A further area where businesses may not fully appreciate the complexity of vendor-supplied maintenance is that traditional methods of costing, cost-benefit analysis (CBA), return on investment (ROI) and risk-analysis, do not translate well from an in-house to a vendor-supplied environment unless specific allowance is made for vendor-supplied maintenance activities. Although a maintenance release may have no compelling reason to be implemented, for example, a low ROI, a negative CBA, no improvement to risk profile, a later more critical maintenance item may have a dependency on the earlier one. If the risk of deferral is not factored into the original implementation decision, the cost and time required to utilize the later critical maintenance release will be under-appreciated.

Traditional budgeting sets an IT department's operating budget on an annual basis. Translation of this budget into staffing allocation extends the assumption of fixed budget into an assumption of staff costs and staff as fixed input, available to perform work. Contained within this work is the effort required to analyze, test, and implement vendor-supplied maintenance into the production environment. However, this study has shown that vendor behavior does not always support such a predictable cycle of resource and budget availability. A case study [20] emphasized that if mandatory, maintenance were implemented when it arrived from the vendor, and 80% of the annual maintenance effort would be consumed through fortnightly implementations; batching updates into larger, less frequent implementation activities significantly benefited the reduction of total effort required. This is an example of planned and managed maintenance deferral. An implication of this somewhat random vendor behavior of maintenance production is to introduce a variable requirement for maintenance work into an organization that is geared for a static level of effort. Incorrectly accounting for this variability through the budgeting cycle could introduce a financial constraint on the ability to implement vendor-supplied maintenance. The proposed lifecycle model with its planning and preparation activities might support this processes.

When surrendering control of maintenance to the vendor, an organization largely relinquishes the ability to manage or dictate the content of an individual maintenance package and thus either might defer as long as possible or see no need for maintenance. Organizations may assume that once purchased no allocation of time or effort is required for ongoing maintenance, and that the problem was solved in the original purchase. This view is reinforced by purchase decisions that fail to build the operational and maintenance costs into the decision process. An alternative outcome from the trigger event may be to re-assess the vendor software and determine that a replacement is necessary as presented in a case for optimally timed system replacement in response to this outcome [44]. In another case an examination of the issues lead to a system being retired, again referencing the maintenance cost/effort of the system to support the decision [10]. In this case, a valid approach is to operate the current system without further maintenance. This is an example of conscious deferral. It is addressed in the proposed maintenance lifecycle wait state with a subsequent trigger event, which results in leaving the maintenance cycle.

7 Conclusions

This study shows, and this is in particular emphasized in work concerning maintenance in security and safety sensitive environments [38], that maintenance behavior in purchasing organizations is not universal, but can be unified into a maintenance lifecycle model that takes different contexts into account. It presents a comprehensive review of the reasons for maintenance deferral and implementation within the area of vendor software from the purchaser's perspective. As such it provides a solid foundation for further attention and research in this long neglected area and demonstrates that execution of a broad systematic literature review relating to a sparsely published area of research informs research through the deduction of themes and concepts as well as a foundational lifecycle model from an expansive selection of literature.

This study supports practice with an understanding of organizational behavior with regard to maintenance deferral and implementation. Through an awareness of common reasons, it can help identify deferral causes, develop responses to these causes and forecast the upcoming need for and implementation of maintenance within an organization. The IS community can advance this process through further research, resulting in theories, frameworks and models that assist practice in navigating the deferral problem in the future. The derived reasons have been distilled from work where they form sometimes an incidental mention during the study of other topics. Though important enough to warrant mentions, the list cannot be considered complete without further empirical testing. Further research is thus needed to validate the identified concepts and themes as well as the Maintenance Lifecycle model and its usefulness for understanding and resolving the maintenance deferral problem with empirical case studies. The differences between reasons leading to the deferral of maintenance and reasons leading to the implementation of maintenance show that understanding the motivations that require an upgrade decision in the present, do not explain all motivations for deferral in the past. Thus a conceptualization of deferral as a process that recognizes that deferment and implementation of maintenance take place in a complex social process may be beneficial in such investigations.

Any study has its limitations. The systematic literature review was performed by the first author without a peer-review and dispute resolution process for the evaluation of each paper against the filtering criteria. The review was limited to papers published in English. Restricting the search to the Web of Science™ exposed the review to constraints of the data source concerning publishing dates, use of search operators and keywords. These limitations may exclude some valid articles. This study has focused on the organizational behaviors relating to single vendor-supplied systems; organizations can operate with multiple, integrated vendor software packages that increase the complexity of any maintenance decision [12]. This has also to be taken into account in future research.

References

1. Horning, J., Neumann, P.G.: Risks of neglecting infrastructure. Commun. ACM **51**(6), 112 (2008)
2. Xu, L., Brinkkemper, S.: Concepts of product software. Eur. J. Inf. Syst. **16**(5), 531–541 (2007)
3. Khoo, H.M., Robey, D., Rao, S.V.: An exploratory study of the impacts of upgrading packaged software: a stakeholder perspective. J. Inf. Technol. **26**(3), 153–169 (2011)
4. Maheshwari, B., Hajnal, C.: Total systems flexibility and vendor developed software: Exploring the challenges of a divided software life cycle, IEEE International Engineering Management Conference 2002, Vols I&II, Proceedings: Managing Technology for the New Economy, IEEE, New York, pp. 218–223 (2002)
5. Carney, D., Hissam, S.A., Plakosh, D.: Complex COTS-based software systems: practical steps for their maintenance. J. Softw. Maint.—Res. Pract. **12**(6), 357–376 (2000)
6. IEEE: IEEE Standard Glossary of Software Engineering Terminology, pp. 1–84 (1990)
7. Swanson, E.B., Chapin, N.: Interview with Swanson, E. Burton. J. Softw. Maint.—Res. Pract. **7**(5), 303–315 (1995)
8. Ng, C.S.P.: A decision framework for enterprise resource planning maintenance and upgrade: A client perspective. J. Softw. Maint. Evolut.—Res. Pract. **13**(3), 431–468 (2001)
9. Khoo, H.M., Robey, D.: Deciding to upgrade packaged software: a comparative case study of motives, contingencies and dependencies". Eur. J. Inf. Syst. **16**(5), 555–567 (2007)
10. Swanson, E.B., Dans, E.: System life expectancy and the maintenance effort: exploring their equilibration. MIS Q. **24**(2), 277–297 (2000)
11. Cusumano, M.A.: Changing software business: moving from products to services. Computer **41**(1), 20–27 (2008)
12. Vigder, M.R., Kark, A.W.: Maintaining COTS-based systems: start with the design. In: Fifth International Conference on Commercial-off-the-Shelf, IEEE Computer Soc, Los Alamitos, 2006, pp. 11–18 (2006)
13. Ng, C.S.P., Gable, G.G., Chan, T. Z.: An ERP-client benefit-oriented maintenance taxonomy. J. Syst. Softw. **64**(2), 87–109 (2002)
14. Harvey, M.: Optimising road maintenance. In: International Transport Forum, OECD/ITF, Paris, 25–26 October 2012, viewed 24 June 2013, http://www.internationaltransportforum. org/jtrc/DiscussionPapers/jtrcpapers.html (2012)

15. Reifer, D.J., Basili, V.R., Boehm, B.W., Clark, B.: Eight lessons learned during COTS-based systems maintenance. IEEE Softw. **20**(5), 94–96 (2003)
16. Webster, J., Watson, R.T.: Analyzing the past to prepare for the future: writing a literature review. MIS Q. **26**(2), 13–23 (2002)
17. Kitchenham, B., Brereton, O.P., Budgen, D., Turner, M., Bailey, J., Linkman, S.: Systematic literature reviews in software engineering—a systematic literature review. Inf. Softw. Technol. **51**(1), 7–15 (2009)
18. Kitchenham, B., Brereton, O.P.: A systematic review of systematic review process research in software engineering. Inf. Softw. Technol. **55**(12), 2049–2075 (2013)
19. Dybå, T., Dingsøyr, T.: Empirical studies of agile software development: a systematic review. Inf. Softw. Technol. **50**(9–10), 833–859 (2008)
20. Ng, C.S.P., Chan, T.Z., Gable, G.G.: A client-benefits oriented taxonomy of ERP maintenance. In: IEEE International Conference on Software Maintenance, Proceedings: Systems and Software Evolution in the Era of the Internet, IEEE Computer Soc, Los Alamitos, 2001, pp. 528–537 (2001)
21. Lientz, B.P.: Issues in software maintenance. Comput. Surv. **15**(3), 271–278 (1983)
22. Khoo, H.M., Chua, C.E.H., Robey, D.: How organizations motivate users to participate in support upgrades of customized packaged software. Inf. Manag. **48**(8), 328–335 (2011)
23. Hybertson, D.W., Ta, A.D., Thomas, W.M.: Maintenance of COTS-intensive software systems. J. Softw. Maint.—Res. Pract. **9**(4), 203–216 (1997)
24. Ketler, K., Turban, E.: Productivity improvements in software maintenance. Int. J. Inf. Manage. **12**(1), 70–82 (1992)
25. Swanson, E.B., Beath, C.M.: Maintaining information systems in organizations. Wiley, New York (1989)
26. Biskup, H., Kautz, K.: Maintenance: nothing else but evolution? Inf. Technol. People **6**(4), 215–231 (1992)
27. Tan, W.G., Gable, G.G.: Attitudes of maintenance personnel towards maintenance work: a comparative analysis. J. Softw. Maint.—Res. Pract. **10**(1), 59–74 (1998)
28. Junio, M.G.A., Malta, N.N., Mossri, H.D., Marques-Neto, H.T., Valente, M.T.: On the Benefits of Planning and Grouping Software Maintenance Requests,15th European Conference on Software Maintenance and Reengineering (CSMR), pp. 55–64 (2011)
29. Hanna, M.L., Martin, L.: Quantitative determination of maintenance concepts for COTS based systems. In: Annual Reliability and Maintainability Symposium, Proceedings, IEEE, New York, 2007, pp. 478–481 (2007)
30. Bloch, H.P.: Deferred maintenance increases pump failures. Power **155**(2), 16 (2011)
31. Anderson, W., McAuley, J.: Commercial off-the-shelf product management lessons learned—satellite ground control system (SGCS) upgrade. In: Fifth International Conference on Commercial-off-the-Shelf, IEEE Computer Soc, Los Alamitos, pp. 206–213 (2006)
32. Lientz, B.P., Swanson, E.B.: Problems in application software maintenance. Commun. ACM **24**(11), 763–769 (1981)
33. Brownsword, L., Oberndorf, T., Sledge, C.A.: Developing new processes for COTS-based systems. IEEE Softw. **17**(4), 48–55 (2000)
34. Gable, G.G., Chan, T.Z., Tan, W.G.: Large packaged application software maintenance: a research framework. J. Softw. Maint. Evolut.—Res. Pract. **13**(6), 351–371 (2001)
35. Bachwani, R., Crameri, O., Bianchini, R., Zwaenepoel, W.: Recommending software upgrades with Mojave. J. Syst. Softw. **96**, 10–23 (2014)
36. Arora, A., Krishnan, R., Telang, R., Yang, Y.B.: An empirical analysis of software vendors' patch release behavior: impact of vulnerability disclosure. Inf. Syst. Res. **21**(1), 115–132 (2010)
37. Ellison, G., Fudenberg, D.: The neo-Luddite's lament: excessive upgrades in the software industry. Rand J. Econ. **31**(2), 253–272 (2000)
38. Arora, A., Forman, C., Nandkumar, A., Telang, R.: Competition and patching of security vulnerabilities: an empirical analysis. Inf. Econ. Policy **22**(2), 164–177 (2010)

39. Gartner: Gartner Estimates Global 'IT Debt' to Be $500 Billion This Year, with Potential to Grow to $1 Trillion by 2015, No. 1, Gartner, gartner.com, p. 1 (2010)
40. Mukherji, N., Rajagopalan, B., Tanniru, M.: A decision support model for optimal timing of investments in information technology upgrades. Decis. Support Syst. **42**(3), 1684–1696 (2006)
41. Ben-Menachem, M.: Towards management of software as assets: a literature review with additional sources. Inf. Softw. Technol. **50**(4), 241–258 (2008)
42. Kaiser, H.H.: Deferred maintenance. New Direct. High. Educ. **30**, 41–54 (1980)
43. Visser, J.K.: Maintenance management—a neglected dimension of engineering management. In: IEEE AFRICON, Vols 1 and 2: Electrotechnological Services for Africa, IEEE, New York, pp. 479–484 (2002)
44. Tan, Y., Mookerjee, V.S.: Comparing uniform and flexible policies for software maintenance and replacement. IEEE Trans. Software Eng. **31**(3), 238–255 (2005)

A Model-Level Mutation Tool to Support the Assessment of the Test Case Quality

Maria Fernanda Granda, Nelly Condori-Fernández, Tanja E.J. Vos and Oscar Pastor

1 Introduction

In Model-Driven Engineering the models or conceptual schemas (CS) are the primary artefacts in the software development process, and efforts are focused on their creation, testing and evolution at different levels of abstraction. If a model has defects, these are passed on to the following stages of the Software Development Life Cycle, including coding. The quality of a CS can be assessed by detecting its defects during execution. The best test suite is the one that has the best chance of finding defects, but how we do know how good a test suite is? Mutation testing is one of the ways of assessing the quality of a test suite. This method injects artificial faults or changes into a CS (mutant generation) and checks whether a test suite is "good enough" to detect these artificial faults. The artificial faults can be created automatically, using a set of mutation operators (MO) to change (i.e. mutate) some parts of the software artefact. Mutants can be classified into two types: First Order

A prior version of this paper has been published in the ISD2016 Proceedings (http://aisel.aisnet.org/isd2014/proceedings2016).

M.F. Granda (✉)
University of Cuenca, Cuenca, Ecuador
e-mail: fernanda.granda@ucuenca.edu.ec

N. Condori-Fernández
Vrije Universiteit van Amsterdam, Amsterdam, The Netherlands
e-mail: n.condori-fernandez@vu.nl

M.F. Granda · T.E.J. Vos · O. Pastor
Universitat Politècnica de València, Valencia, Spain
e-mail: tvos@pros.upv.es

O. Pastor
e-mail: opastor@pros.upv.es

© Springer International Publishing Switzerland 2017
J. Gołuchowski et al. (eds.), *Complexity in Information Systems Development*,
Lecture Notes in Information Systems and Organisation 22,
DOI 10.1007/978-3-319-52593-8_2

Mutants (FOM) and Higher Order Mutants (HOM) [1]. FOMs are generated by applying mutation operators only once. HOMs are generated by applying mutation operators more than once [2]. However, approaches that employ mutation testing at higher levels of abstraction, especially on CS, are not common [2].

One problem in the design of tests to assess test case quality is that real software artefacts of appropriate size including real faults are hard to find and hard to prepare appropriately (for instance, by preparing correct and faulty versions) [3]. Even when software artefacts with real faults are available, these faults are not usually numerous enough to allow the experimental results to achieve statistical significance [3]. Thus, mutation testing is usually considered expensive due to: (i) the large number of mutants generated; (ii) the time-consuming task of determining equivalent mutants (i.e. functionally identical to the original artefact although syntactically different); and (iii) the time required to compile and execute the mutants [4]. This means mutation testing of real-world software would be extremely difficult without a reliable, fast and automated tool that: (a) generates mutants, (b) runs the mutants against a test suite and (c) reports the mutation score of the test suite.

This paper describes a mutation tool that generates FOMs for CS based on UML Class Diagram (CD) by using previously defined mutation operators [5]. The main usefulness of the mutation tool is to support a well-defined, fault-injecting process to assess the test case quality at the CS level.

The novel contributions of this paper are: (1) the MµtUML prototype mutation tool designed to generate FOMs for UML CD-based CS, eliciting its benefits and weaknesses. (2) An evaluation of the effectiveness and efficiency of the mutation tool to generate valid and non-equivalent FOMs of UML CD-based CS by using six subject CSs.

The rest of this paper is organized as follows. Section 2 describes the background to the study and Sect. 3 describes the mutation tool itself. The empirical evaluation is described in Sect. 4. Section 5 presents the results of the evaluation by applying 18 mutation operators to six CSs and a discussion on effectiveness and efficiency of the proposed mutation tool. Section 6 describes possible threats to validity. Section 7 summarizes our conclusions and outlines future work.

2 Background

2.1 *Executable Conceptual Schema Based on UML Class Diagram*

In this paper, defects will be introduced by deliberately changing a UML CD-based CS, resulting in wrong behaviour and possibly causing a failure. As the CS of a system should describe its structure and behaviour (constraints), we represent it by a UML-based (CD). A class diagram is the UML's main building block and shows elements of the system at an abstract level (e.g. class, association class), their properties (owned attributes), relationships (e.g. association and generalization) and operations. In a UML, operations are specified by defining pre- and post-conditions (i.e. constraints) [6]. In this paper we evaluate mutation operators that can inject defects into the following elements: class, attribute, operations, parameters, associations and constraints. In this context, an executable UML model is one with a behavioural specification detailed enough to effectively be run as a program. There are several model execution tools and environments.[1] However, each tool defines its own semantics for model execution, often including a proprietary action language, and models developed in one tool could not be interchanged with or interoperated with models developed in another tool.

In this work, we use the action language adopted as a standard by OMG,[2] which is known as the Action Language for Foundational UML, or Alf [7], which is basically a textual notation for UML behaviours that can be attached to a UML model at any point where there is UML behaviour, e.g. the method of an operation or the classifier behaviour of a class. As Alf notation includes basic structural modelling constructs, it is also possible to do entire models textually in Alf. Semantically, Alf maps the model to the Foundational UML (fUML [8]) subset, after which fUML provides the virtual machine for the execution of the Alf language.

2.2 Mutant Generation Time Estimation Model

The usual process for obtaining values of time on task data involves recruiting users and then performing tests with them in a lab. This procedure, while providing a wealth of informative data can be expensive and time-consuming [9].

Since one of the goals of this study was to analyse the time saved in the mutant generation process by using the proposed tool, we required a method that measured experienced-user task time in order to estimate the time required to generate each mutant type analysed. The most familiar of these cognitive modelling techniques is GOMS (Goals, Operators, Methods and Selection Rules), which has been documented in the still highly referenced text "The Psychology of Human Computer Interaction", by Card, Moran and Newell (1983) [10]. GOMS represents a family of techniques, the most familiar of which is Keystroke-Level Modelling [10]. We selected the Keystroke-Level Model for this study because it has revealed remarkably precise prediction results in several projects such as [11, 12]. The Keystroke-Level Model predicts the task execution time of a specified interface and task scenario. Basically, it requires a sequence of keystroke-level actions the user must perform to accomplish a task and then adds up the total time required for the

---

[1]http://modeling-languages.com/list-of-executable-uml-tools/.

[2]http://www.omg.org/.

actions. The actions are termed at keystroke level if they are actions like pressing keys, moving the mouse, pressing buttons, and so on [13]. The values used for this technique are described in detail in Sect. 4.2.

3 MutUML: A Mutation Tool

The most critical activity in mutation testing is the suitable design of mutation operators so that they reflect typical defects of the artefact under test. In a previous work [14], we presented a defects classification at model level and in [5] described the process of selection of the 18 mutation operators from a list of 50 for generating First Order Mutants to UML CD-based CS.

We developed a mutation tool (https://staq.dsic.upv.es/webstaq/mutuml.html) for generating first order mutants by using a set of 18 previously defined mutation operators [5], which specify the changes and restrictions required for each mutation operator (see Table 1).

MμtUML provides a graphical user interface as shown in Fig. 1.

Table 1 Mutation Operators for FOMs taken from [5]

#	Code	Mutation operator description
1	UPA2	Adds an extraneous Parameter to an Operation
2	WCO1	Changes the constraint by deleting the references to a class Attribute
3	WCO3	Change the constraint by deleting the calls to specific operation
4	WCO4	Changes an arithmetic operator for another and supports binary operators: +, −, *, /
5	WCO5	Changes the constraint by adding the conditional operator "not"
6	WCO6	Changes a conditional operator for another and supports operators: or, and
7	WCO7	Changes the constraint by deleting the conditional operator "not"
8	WCO8	Changes a relational operator for another operators: <, <=, >, >=, ==, !=
9	WCO9	Changes a constraint by deleting a unary arithmetic operator (−)
10	WAS1	Interchange the members (memberEnd) of an Association
11	WAS2	Changes the association type (i.e. normal, composite)
12	WAS3	Changes the memberEnd multiplicity of an Association (i.e. *−*, 0..1−0..1, * −0..1)
13	WCL1	Changes visibility kind of the Class (i.e. private)
14	WOP2	Changes the visibility kind of an operation
15	WPA	Changes the Parameter data type (i.e. String, Integer, Boolean, Date, Real)
16	MCO	Deletes a constraint (i.e. pre-condition, post-condition constraint, body constraint)
17	MAS	Deletes an Association
18	MPA	Deletes a Parameter from an Operation

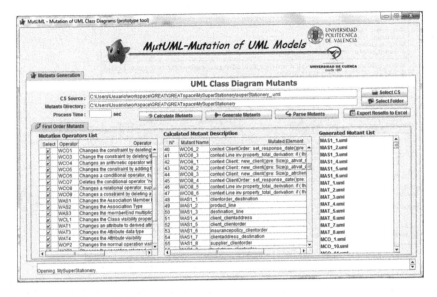

Fig. 1 MutUML screenshot

The tool functionality is separated into the following three processes:

1. Calculating Mutants. Testers can select the CS source file (.uml) to calculate the FOMs and also the mutation operators to apply (by default all mutation operators are selected). On pressing the "Calculate Mutants" button, the tool calculates the mutants by applying the mutation operators. The information for each mutant is shown in the "Mutant Description Table" and can be exported as a report by pressing the "Export Report to Excel" button.

2. Generating Mutants. The testers/designers can create the mutants required by selecting from the previously calculated mutant list (by default all mutants are selected) and pressing the "Generate the Mutants" button to generate them. The tool generates the CS mutants (.uml) from the CS source file (.uml).

Parsing Mutants. After the mutants have been generated they need to be analysed by the parser. This analysis is required before the mutation testing process and also to automatically classify the mutants as valid or non-valid. Each mutant is transformed into an executable format by using Alf language. The Alf parser produces an output with the analysis results of each mutant. To understand how MμtUML works we refer to the partial view of a CS in Fig. 2. Five mutation operators have been applied to the CS. Four operators generate valid FOM [i.e. (b) UPA2, (c) WAS3, (d) WCO3, (e) MCO]. However, applying the MAS operator to the *WhiteCells* association generates a non-valid FOM because there is a constraint (i.e. MovieUnique) that is related with the association. Simply deleting the association would result in a Dangling constraint, which evidently is

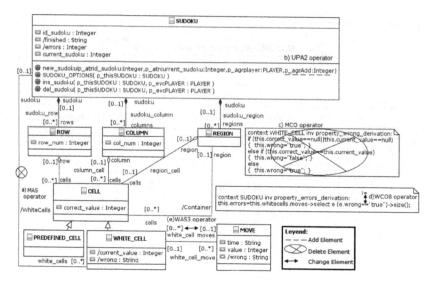

Fig. 2 Excerpt of a UML CD-based CS and the application of five mutation operators, adapted from [15]

not desirable. Therefore, we need to add more steps to the operator (going from FOM to HOM). The HOM should delete the association together with the respective constraint. This way, the mutant will not be detected by the parser and can generate a valid mutant for testing.

4 Empirical Evaluation

This section describes the goal, the research questions, the metrics, the evaluation context and procedure followed for the evaluation.

4.1 Goal

In accordance with the Goal/Question/Metric Paradigm [16] the goal of our empirical study is as follows: **To analyse** the mutant generation strategy of the Mutation tool, **for the purpose of** carrying out an evaluation **with respect to** the effectiveness and efficiency in generating valid First Order Mutants to UML CD-based CS **from the viewpoint of** the researchers.

4.2 Research Questions and Metrics

By means of this study, we aim to be able to respond to the following research questions (RQ):

RQ1: How effective are the mutation operators implemented in a mutation tool for generating FOMs of Conceptual Schema?

As this RQ is focused on the generation strategy, the following research questions and metrics are derived from it:

1. RQ1.1. For each defined mutation operator, what is the percentage of valid mutants generated by the mutation tool? The metric M1 for RQ1.1 is the percentage calculated by dividing the number of valid mutants generated by the tool by the total number of mutants that can be generated from the CS elements. The number of mutants that can be generated determines the cost of creating and executing them and also the cost of deriving test cases that kill them. The number of non-valid mutants has an impact on the cost of identifying and discarding them. The mutation tool can indicate whether a mutant is valid or not according to the restrictions defined for each mutation operator.

$$M1(MO) = \frac{M_V(MO)}{M_G(MO)} \times 100\% \qquad (1)$$

2. RQ1.2. For each defined MO, what percentage of parsed mutants is equivalent? The first metric M2 for RQ1.2 is the percentage calculated by dividing the number of valid mutants that are equivalent by the number of valid mutants for mutation operator. The number of equivalent mutants has an impact on the cost of performing mutation testing because a tester needs to execute the test cases against the equivalent mutants to identify and discard them.

$$M2(MO) = \frac{M_E(MO)}{M_V(MO)} \times 100\% \qquad (2)$$

3. The second metric M3 for RQ1.2 is the percentage calculated by dividing the number of equivalent mutants that can be eliminated using the proposed tool, and total number of equivalent mutants generated by an operator. The cost of performing mutation testing of equivalent mutants can be reduced by the tool by automating an analysis of the subject CS to identify them.

$$M3(MO) = \frac{M_E(MO) \, detected \, by \, M\mu tUML}{M_E(MO)} \times 100\% \qquad (3)$$

RQ2: To what extent is the generation time reduced by using the tool?

The metric M4 for this RQ is the percentage of time saved by the mutation tool when generating mutants (FOMs). The Manual Generation time is measured by the generation time of valid and non-valid mutants for the subject CS. While the Tool Generation time is calculated by adding the times required to calculate, generate and parse the mutants (FOMs) when using the MμtUML tool. This metric can be measured by applying the following formula:

$$M4(CS) = \frac{Manual\ Generation\ Time\ (CS) - Tool\ Generation\ Time\ (CS)}{Manual\ Generation\ Time\ (CS)} \times 100\%$$

$$(4)$$

In order to predict the times for manual generation, the following step-by-step description adapted from Kieras [13] was used to apply the Keystroke-Level Model method in this work.

1. Choose a representative task scenario for each mutation operator. The general scenario required to create manually a CS mutant is: (a) Task 1—open the CS source file, (b) Task 2—duplicate the CS source file, (c) Task 3—select the CS element, (d) Task 4—apply the mutation operator, this task is particular for each mutation operator (see Table 3), and (e) Task 5—save the mutant and close it.
2. List the keystroke-level actions involved in doing each task with the execution times. The following are some of the standard keystroke-level actions and estimated times for each operator [13].

   - M: Mental operation: User decides or reflects where to click (1.2 s)
   - H: Home: User moves hand between keyboard and mouse (0.4 s)
   - P: Point: User point with the mouse to a target on the screen (1.1 s)
   - K: Set: User clicks on the target (0.28 s). The time considers the average non-secretarial typist (40 wpm—words per minute) [9].
   - B: Button: User clicks on the button (0.1 s).
   - BB: User pushes and releases the mouse button rapidly, as in a selection click (0.2 s).

As we assumed waiting time to be negligible we did not deal with any physical operators for it.

3. List and calculate time for each composite action. Using the granular steps from Keystroke-Level Model, a composite action is clicking on a menu option of the UML diagram editor such as File/Open, Save and Save as, so the four steps are replaced with the composite action: Click on Option. The time to complete this action is modelled as: M (1.2) + P (1.1) + H (0.4) + B (0.1) = approximately 2.8 s. Using this method, we defined a small number of composite actions to account for almost all the above five tasks in the 18 mutation operators. The composite actions used were as follows:

- CA1: Click an Option/Button (MPHB = 2.8 s).
- CA2: Double click (MPHBB = 2.9 s).
- CA3: Typing Mutant name in a Text Field of the Dialog box. The mutant name is formed by the code of mutant (3 uppercase letters) + an underscore ("_") + a sequential number formed by 3 digits + the file extension ".uml-class" (18 K = 5.32 s). Pressing the SHIFT key counts as a separate keystroke.
- CA4: Pull-Down List (3.04) (time taken from [9]).
- CA5: Scrolling (3.96 s) (time taken from [9]).

4. Estimate the time to complete all scenarios for each mutation operator. For this study we assumed that the person creating the mutants was skilled in: (a) modelling UML CD-based CS, (b) using the UML CD editor (e.g. UML2 tool), and (c) applying the different 18 mutation operators. It was also assumed that the FOMs list had been calculated previously, the UML CD editor had been loaded and active and finally that the tasks were error-free. The Keystroke-Level Model thus addressed only a single aspect of task performance and did not consider other dimensions, such as error-free execution, concentration, fatigue and so on [10]. Using the defined composite action times, the times for each task were calculated (see Tables 7 and 8 in Appendix 1).

Table 7 shows the times in seconds estimated by the method for tasks common (i.e. Tasks 1, 2, 3, and 5) to all mutation operators. Table 8 shows the sequence of composite actions required for each mutation operator for Task 4 and the time predicted by Keystroke-Level Model for this task. The time for each mutation operator was estimated in the last column of Table 8 by using the times of the respective tasks for each operator (Task 1–Task 5).

4.3 Evaluation Context

Subject CSs.

Six subject CSs were used in the study (see elements in Table 2). These contained a variety of possible characteristics present in UML CD-based CS, including classes, relationships (i.e. association, composite aggregation, and generalization) and different types of constraints (i.e. pre-condition, post-condition and body condition). Some were found in the literature (i.e. [15, 17, 18]) and others were selected because they contained the CS elements required to inject the faults.

A brief description of each CS is as follows:

1. The Medical Treatment (MT) CS defines part of the CS (of a Medical Treatment business process) of a fictional hospital named University Hospital Santiago Grisolía, developed by España et al. [17].
2. The Sudoku Game (SG) CS was developed by Tort and Olivé [15] as an object-oriented CS of the Sudoku Game system and this CS defines the

Table 2 Elements of the subject Conceptual Schemas

Element	MT	SG	ER	OCR	SS	PA
Classes	6	11	7	10	9	15
Attributes	26	26	36	61	44	43
Derived attributes	0	6	6	1	1	33
Operations	13	19	24	16	32	30
Parameters	43	48	75	77	91	82
Associations	5	6	8	10	9	19
Derived associations	0	2	0	0	0	0
Composite aggregations	0	3	0	0	0	0
Constraints	9	19	21	14	12	45
Generalizations	0	4	0	3	0	0

functionality for managing different users, who play with Sudokus and generating new games.

3. The Expense Report (ER) CS defines the functionality of an information system to manage the expense report life cycle of a business and deals with several entities such as departments, employees, projects and expense types.

4. The Online Conference Review (OCR) CS, which is based on the description of the CyberChair System, defines the functionality of an information system to deal with members (committee chair and program committee) of a conference, as well as authors that submit papers to a conference to be evaluated for acceptance.

5. The Super Stationery (SS) CS defines the information system of a company that provides stationery and office material to its clients. This CS was developed by España et al. [18].

6. The Photography Agency (PA) CS defines the information system that manages photographers and their photographic reports for distribution to newspaper publishers.

Tools.

In this paper one of the aims is to predict the efficiency of the mutation tool in generating valid and non-equivalent mutants in a UML CD editor. The Keystroke Level Model (KLM) Calculator (http://courses.csail.mit.edu/6.831/2009/handouts/ac18-predictive-evaluation/klm.shtml) is used for calculating the predictions of task execution times in the UML CD editor from defined scenarios for applying the different mutation operators. This choice is motivated by the large number of publications in the Computer Human Interaction environment using KLM in a variety of emerging application domains [19].

On the other hand, there is no literature available on tools with both integrated functionality (i) the automated generation of test cases and (ii) tests execution for Conceptual Schemas. Therefore, we used our CoSTest CS testing tool (https://staq.dsic.upv.es/webstaq/costest.html) for this work. This tool generates test cases by applying a Model-Driven approach [20]. The test cases use assertions on the return values of the methods and compare them with the post-conditions. We performed the following steps to use the CoSTest tool with UML CD-based CS:

- For each CS, we set the CS testing tool to generate the test cases. We provided the CS testing tool with a requirements model and a set of input values suitable for the subject CS, after which the tool generated the test suite.
- Since a CS is not designed for the CS testing tool, the tool generates an executable CS by applying a transformation from UML to the Alf language.
- The test cases generated by the tool were executed against the CS under test by using the virtual machine for the execution of the Alf language.

A full description of the testing tool is beyond the scope of the present paper.

Finally, while there are tools that support manipulation of UML-based Conceptual Schemas such as Papyrus (http://www.eclipse.org/papyrus/) and UML2 Tools (http://wiki.eclipse.org/MDT-UML2Tools). The UML2 tool is an Eclipse Modelling Framework-based implementation, which is integrated into the tool used for modelling the requirements used as input in the CoSTest tool. For this reason we selected this tool for manipulating UML models.

4.4 Procedure

With the aim of finding empirical evidence to answer the aforementioned RQs, we divided our study into two parts:

The first evaluation was performed to answer RQ1 and partially answer RQ2. In the first evaluation we generated mutants for each mutation operator from the subject CS, then identified non-valid mutants by applying the MO restrictions and provided suggestions on how to reduce the percentage of equivalent mutants. These results are given in Sect. 5.1.

The second evaluation was designed to answer RQ2. For this we derived a reliable way to estimate time-on task by using the metric defined in Sect. 4.2 and the tool summarized in Sect. 4.3. The generation time savings when using the tool for each subject are given in Sect. 5.2.

5 Results Analysis

5.1 Effectiveness: Mutant Generation from FOMs Mutation Operators

Table 3 summarizes the results of generating and parsing the mutants for the subject CS by using the mutation tool.

Table 3 Generated and valid mutants using mutation tool

MO	CS		SG		ER		OCR		SS		PA		ALL		M1 (%)	C (%)
	MT															
	M_F	M_G	M_F	M_G	M_F	M_G	M_F	M_G	M_F	M_G	M_F	M_G	M_F	M_G		
UPA2	13	13	19	19	24	24	16	16	32	32	30	30	134	134	100	100
WCO1	0	N/A	7	7	9	9	1	1	3	3	33	33	53	53	100	100
WCO3	13	0	19	0	29	5	16	0	34	2	43	13	154	20	100	100
WCO4	0	N/A	15	15	8	8	0	N/A	2	2	26	26	51	51	100	100
WCO5	0	N/A	11	11	11	1	6	6	2	2	5	5	35	35	100	100
WCO6	0	N/A	12	12	2	2	5	5	3	3	4	4	26	26	100	100
WCO7	0	N/A	1	1	0	N/A	0	N/A	0	N/A	0	N/A	1	1	100	100
WCO8	6	6	47	47	21	21	26	26	13	13	34	34	147	147	100	100
WCO9	0	N/A	1	1	0	N/A	0	N/A	0	N/A	0	N/A	1	1	100	100
WAS1	5	4	11	0	8	0	10	7	9	6	19	5	62	22	100	35.5
WAS2	5	5	11	11	8	8	10	10	9	9	19	19	62	62	100	100
WAS3	15	12	33	0	24	0	30	21	27	18	57	15	186	66	100	35.5
WCL1	6	6	11	11	7	7	10	10	9	9	15	15	58	58	100	100
WOP2	13	13	19	19	24	24	16	16	32	32	30	30	134	134	100	100
WPA	43	9	48	9	75	17	77	3	91	26	82	12	416	76	100	18.3
MCO	9	9	19	11	21	15	14	13	12	11	45	12	120	71	100	59.2
MAS	5	4	11	0	8	0	10	7	9	6	19	5	62	22	100	35.5
MPA	43	10	48	11	75	23	77	6	91	32	82	18	416	101	100	24
All	176	91	343	185	354	175	324	147	378	206	543	276	2118	1079	100	50.9

For each mutation operator in Table 3 we show the number of (valid and non-valid) possible mutants (MP) that can be generated from the CS elements, the number of mutants generated by the tool (MG) for each of the subject CS, as well as the total number of possible generated mutants and the total number of mutants generated by the tool for all CS. Column M1 shows the percentage of valid mutants generated by the tool from all CS. The highest percentage of M1 (100%) was achieved by implementing the rules and restrictions of the FMOs. It can be seen that the total number of non-valid mutants (1039 or 49.1%) is lower than the valid mutants (1079 or 50.9%). The last cell of the last column (C%) in Table 3 shows the percentage of valid mutants for each mutation by applying the restrictions of the FOMs for all six subject CSs.

We manually analysed the mutants to determine whether they were equivalent (i.e. the CS mutant produces the same output as the original CS as if it had no faults). The analysed output is produced by the CS testing tool. An example of an equivalent mutant is shown in Fig. 3, where the mutation is not detected by the CS testing tool.

Table 4 shows the results of analysing equivalent mutants generated in the six CS. For each CS, the table shows the number of equivalent mutants and the percentage of equivalent mutants out of the valid mutants generated by each operator. For example, the first row in Table 4 shows that operator WCO4 had 2 equivalent mutants in the Sudoku CS. These contribute about 13.3% of the 15 valid mutants that the operator has generated. Column M2 in Table 4 shows the percentage of equivalent mutants of the total number of valid mutants for each operator. The last column in Tables 4 shows the percentage of equivalent mutants generated by each operator out of the total number of equivalent mutants. For example, in the first row of Table 4, the WCO4 operator generated 2 equivalent mutants out of the 78 equivalent mutants that the mutation tool generated in all the CS. It therefore, contributed about 2.6% of the total number of equivalent mutants. The last column in the table shows that most of the equivalent mutants were generated by the WOP2 operator with 74.3% of the total number of equivalent mutants.

We inspected the equivalent mutants to determine why the mutants generated cannot be detected. The reason is that the WOP2 operator (changes the operation visibility) when it is applied on a constructor operation, only affects the access inherited by child classes (a private constructor of the super class is not inheritable). Therefore, it is impossible to detect this mutation operator when the operation is executed in the test cases. A restriction in the rule of the WOP2 mutation operator should be included in the tool to avoid generating this type of mutant. There are

```
Context WHITE_CELL inv property_current_value_derivation:
// Original Constraint with Relational Operator "=="
this.current_value=this.moves->size()==0?-1:this.current_value= this.moves->last().value;
// Mutant Constraint with Relational Operator "<="
this.current_value=this.moves->size()<=0?-1:this.current_value= this.moves->last().value;
```

Fig. 3 Excerpt of a Constraint mutated by WCO8 operator

Table 4 Number and percentage of equivalent mutants generated using the mutation tool

MO	CS															
	MT		SG		ER		OCR		SS		PA		ALL			
	ME	%	ME	%	ME	%	ME	%	ME	%	ME	%	ME	M2 (%)	C (%)	M3%
WCO4			2	13.3					1				2	3.9	2.6	0
WCO6			1	8.3						33.3			2	7.7	2.6	0
WCO8			6	12.8	1	4.8	3	11.5			6	17.6	16	10.9	20.5	0
WOP2	6	46.2	11	57.9	7	29.2	10	62.5	9	28.1	15	50	58	43.6	74.3	74.3
All	6	6.6	2	10.8	8	4.6	13	8.8	10	4.9	21	7.6	78	7.2	100	74.3

other equivalent mutants such as WCO4, WCO6 and WCO8, which can only be identified by inspecting the mutants. The mutation tool cannot avoid producing them. However, by including the above-described implementation restriction for operator WOP2, we see that 74.3% (Metric M3) of the equivalent mutants generated by the mutation tool can be eliminated.

5.2 Efficiency: Generation Time Reduction by Using the Mutation Tool

We estimated the time of manual generation of valid and non-valid mutants by using the calculated times for each mutation operator in Sect. 4.2 (see Table 3).

Table 5 summarizes the times obtained for each mutation operator in each subject CS by generating valid first order mutants.

The results show that the subject MT has the lowest mutation time (3661.2 s) and subject PA the highest (12516.2 s). The last column in Table 5 shows that most time is required to create mutants by using the WCO8 mutation operator (6894.3 s), and the shortest time is required to create the mutants by using the WCO7 and WCO9 operators (46.1 s). These results are as expected, because these operators

Table 5 Reduced time in generating valid mutant by using the mutation tool

MO	MT (s)	SG (s)	ER (s)	OCR (s)	SS (s)	PA (s)	All (s)
UPA2	520.0	760.0	960.0	640.0	1280.0	1200.0	5360.0
WCO1		413.3	531.4	59.0	177.1	1948.3	3129.1
WCO3	1018.2		391.6		156.6	1018.2	1566.4
WCO4		695.1	370.7		92.7	1204.8	2363.3
WCO5		475.9	475.9	259.6	86.5	216.3	1514.1
WCO6		596.4	99.4	248.5	149.1	198.8	1292.2
WCO7		46.1					46.1
WCO8	281.4	2204.3	984.9	1219.4	609.7	1594.6	**6894.3**
WCO9		46.1					46.1
WAS1	185.0			323.8	277.6	231.3	1017.7
WAS2	188.1	413.8	301.0	376.2	338.6	714.8	2332.4
WAS3	562.8			984.9	844.2	703.5	3095.4
WCL1	232.7	426.6	271.5	387.8	349.0	581.7	2249.2
WOP2	504.1	736.8	930.7	620.5	1241.0	1163.4	5196.5
WPA	447.7	447.7	845.6	149.2	1293.2	596.9	3780.2
MCO	263.3	321.9	438.9	380.4	321.9	351.1	2077.5
MAS	117.0			204.8	175.6	146.3	643.7
MPA	4359.0	394.9	825.7	215.4	1148.8	646.2	3590.0
All	**3661.2**	7978.7	7427.2	6069.5	8541.5	**12516.2**	46194.3

generated the highest and lowest values in the number of valid mutants in the six CSs. Some fields in Table 5 are empty because the different subject CS had not the required elements by these mutation operators.

Figure 4 shows the total times that are reduced by avoiding the manual generation of valid (i.e. 46194.3 s) and non-valid mutants (i.e. 48833.4 s) in the six subject CSs.

From this chart we can see that the generation time depends on number of mutants and the time required by each applied MO. For example, in subject MT, the number of valid mutants (91) is higher than non-valid mutants (85) (see Table 3). However, the generation time of non-valid mutant is the longer time (see Fig. 4). This result is because the some mutation operators for generating non-valid mutants (e.g. WCO3, WPA) require a longer time (see Tables 3 and 8).

Additionally, we estimated the time needed for manually creating equivalent mutants by using the WOP2 operator (58 equivalent mutants * 38.78 s/WOP2 mutant = 2249.24 s), which can be avoided by implementing the suggestion described in Sect. 5.1 in the mutation tool.

Finally, Table 6 shows the time required to calculate, generate and parse the mutants by using our mutation tool for the different subject CS in this study. The results show that the calculation time is negligible when using the tool, while manual generation time is the longest. The time percentage can thus be reduced by more than 92% in the six evaluated CSs by using the proposed mutation tool.

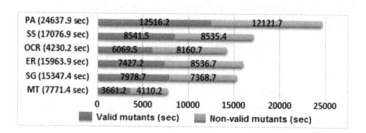

Fig. 4 Time required by a manual generation of valid and non-valid mutants

Table 6 Time used by the mutation tool for generating mutants (FOM) in the subject CS

Task Time	MT	SG	ER	OCR	SS	PA
Using the tool (s)	527.05	1165.10	1053.09	585.07	808.07	1506.16
Reduced Time (s)	7244.35	14182.30	14910.81	13645.13	16268.83	23131.74
Reduced % (M4)	93.2%	92.4%	93.4%	95.9%	95.3%	93.9%

6   Threats to Validity and Limitations

There are several threats that potentially affect the validity of our study including threats to internal validity, threats to external validity, and threats to construct validity.

Threats to internal validity are conditions that can affect the dependent variables of the experiment without the researcher's knowledge. In our study, the selection of mutation operators is the main threat to internal validity. In order to minimize this threat we used a set of 18 previously defined mutation operators [5] to inject faults systematically.

Threats to external validity are conditions that limit the ability to generalize the results of our experiments to industrial practice. This threat is reduced by using six CSs of different sizes (see Sect. 4.2) and domains (e.g. information systems, games). Some well-documented CS were found in the literature (i.e. [15, 17, 18]), and others (i.e. ER, OCR and PA) were selected because they contained the relevant CS elements required to inject the faults.

Threats to construct validity refer to the suitability of our evaluation metrics. We used well-known metrics to measure the effectiveness (number of valid and equivalent mutants) and efficiency (time needed to generate the mutants). In order to perform a specific quantitative analysis for the time saved in the FOM generation process by using the mutation tool. We estimated the time that a user needs to perform the task manually using the Keystroke-Level Model [10]. This model appears to us simple, accurate, and flexible enough to be applied in evaluation situations like ours. However, the model addresses only a single aspect of task performance and does not consider other dimensions, such as error-free execution, concentration, fatigue and so on [10]. Also, this model has several restrictions (e.g. the user must be an expert; the method must be specified in detail; and the performance must be error-free). However, we believe that this model represents an appropriate estimation of the time saved by using the mutation tool and that there is hence little threat to the construct validity.

7   Conclusions and Future Work

Mutations applied at the model level can improve early development of high quality test suites and can contribute to developing high quality systems, especially in a model-driven context. In this paper, we propose a tool that automates the generation of mutants for UML CD-based CS by using a set of previously defined mutation operators. This tool was evaluated for its effectiveness and efficiency in terms of its percentage of valid and non-equivalent mutants and the time that can be saved by using it.

The results show that the mutation operators can be automated avoiding the generation of a high percentage (49.1%) of non-valid mutants. Thus, the tool generates a low percentage (7.2%) of equivalent mutants. However, detecting these mutants is costly in terms of the time and effort of creating, executing and manually inspecting them. We therefore implemented the restrictions and rules for eliminating them by performing a static analysis of the CS. As these results show, the reduction achieved in this analysis of equivalent mutants is about 74.3%, which is equivalent to 2249.24 s estimated by KLM, and the cost of reducing non-valid mutant is 49.1% (48833.4 s estimated by KLM) by using the mutation tool in the six subject CSs involved in this study. Therefore, the results of this study suggest that the mutation tool can help researchers and supports a well-defined, fault-injecting process to generate a potentially large number of valid and non-equivalent FOMs, increasing the statistical significance of results obtained in assessing test case quality.

This study is a part of a more extensive research project, whose main goal is to propose an approach for testing-based conceptual schema validation in a Model-Driven Environment. We have identified three directions in which to extend this work. First, we intend to study the use of HOMs and subsuming HOMs for UML CD-based CS in order to cover all CS elements and other types of faults in UML CD-based models. Secondly, we hope to evaluate the use of HOMs and compare them with FOMs in order to reduce the cost of mutation analysis. Finally, we plan to perform a large-scale empirical study on several industrial subject CS in order to evaluate the effectiveness of the automatized mutation operators in the mutation tool.

Acknowledgements This work has been developed with the financial support by SENESCYT of the Republic of Ecuador, SHIP (SMEs and HEIs in Innovation Partnerships, ref: EACEA/A2/UHB/CL 554187), PERTEST (TIN2013-46928-C3-1-R), European Commission (CaaS project) and Generalitat Valenciana (PROMETEOII/2014/039).

Appendix 1

See Tables 7 and 8.

Table 7 Estimated Time by Keystroke-Level Model

Tasks	Operator sequence	Time (s)
Task 1	CA1 (Open option)	2.9
Task 2	CA1 (File option) + CA1 (Save As option) + CA3 + CA1 (Ok button)	13.72
Task 3	CA5 + CA1	6.76
Task 5	CA1 (Save option) + CA1 (Close button)	5.6
All four tasks		28.98

Table 8 KLM Estimation for Task 4 by each Mutation Operator

MO	Task 4 Scenarios for each mutation operator	Task 4 time (s)	Tasks 1–5 time (s)
UPA2	CA2 (select operation) + CA1 (locate the place to edit the parameter) + 18K (Parameter: String) + 1 K <enter>	11.02	40.00
WCO1	CA1 (body property) + CA1 (edit button) + CA1 (remove button) + CA2 (select attribute) + K (<supr> press) + 4 K (<shift>) + K ("_") + 4 K (var_auxi) + CA1() + 4 K (<shift>) + K ("_") + 5 K (<shift>) + 6 K (<shift>) 5 K (<shift>) + 7 K (<shift>) + K + CA1 (add button) + CA1 (Ok button)	30.6	59.04
WCO3	CA2 (select constraint) + CA1 (locate the place to edit a variable) + 23 K (e.g. var_auxi = new Real (0,0);) + CA1 (locate the place to replace the operation for variable) + CA2 (select operation) + 8 K (var_auxi) +1 K <enter>	20.36	78.32
WCO4	CA1 (body property) + CA1 (edit button) + CA1 (remove button) + CA1 (text point) + K (<supr> press) + K (write operator) + CA1 (add button) + CA1 (OK)	17.36	46.34
WCO5	CA1 (body property) + CA1 (edit button) + CA1 (remove button) + CA1 (text point) + K (write operator) + CA1 (add button) + CA1 (OK)	14.28	43.26
WCO6	CA1 (body property) + CA1 (edit button) + CA1 (remove button) + CA1 (text point) + K (<supr> press) + K (<supr> press) + K (write operator) + K (write operator) + CA1 (add button) + CA1 (OK)	19	49.70
WCO7	CA1 (body property) + CA1 (edit button) + CA1 (remove button) + CA1 (text point) + K (<supr> press) + CA1 (add button) + CA1 (OK)	17.08	46.06
WCO8	CA1 (body property) + CA1 (edit button) + CA1 (remove button) + CA1 (text point) + K (<supr> press) + K (<supr> press) + K (write operator) + K (write operator) + CA1 (add button) + CA1 (OK)	17.92	46.90
WCO9	CA1 (body property) + CA1 (edit button) + CA1 (remove button) + CA1 (text point) + K (<supr> press) + CA1 (add button) + CA1 (OK button)	17.08	46.06
WAS1	CA1 (Source End) + CA1 (Type property) + CA4 (select type) + CA1 (Target End) + CA1 (Type property) + CA4 (select type)	17.28	46.26
WAS2	CA1 (Target End) + CA1 (Aggregation property) + CA4 (select aggregation type)	8.64	37.62
WAS3	CA1 (Source End) + CA1 (select Lower) + K (new value) + CA1 (select Upper) + K (new value) + CA1 (Target End) + CA1 (select Lower) + K (new value) + CA1 (select Upper) + K (new value)	19	46.90

(continued)

Table 8 (continued)

MO	Task 4 Scenarios for each mutation operator	Task 4 time (s)	Tasks 1–5 time (s)
WCL1	CA5 (Class properties) + CA1 (Visibility property) + CA4 (select visibility)	9.8	38.78
WOP2	CA5 (Operation properties) + CA1 (Visibility property) + CA4 (select visibility)	9.98	38.78
WPA	CA1 (right button) + CA1 (parameters manage) + CA1 (select data type) + CA1 (select edit button) + CA5 (data types) + CA1 (OK button) + CA1 (OK button of parameters manage)	21.84	49.74
MCO	K (<supr> press)	0.28	29.26
MAS	K (<supr> press)	0.28	29.26
MPA	K (F2 to edit operation) + CA2 (select the attribute) + k (<supr> press) + CA2 (select the data type) + k (<supr> press) + k (<supr> press on ":")	6.92	35.90

References

1. Jia, Y., Harman, M.: Higher order mutation testing. Inf. Softw. Technol. **51**, 1379–1393 (2009)
2. Jia, Y., Harman, M.: An analysis and survey of the development of mutation testing. Softw. Eng. IEEE Trans. **37**, 1–31 (2011)
3. Andrews, J.H., Briand, L.C., Labiche, Y.: Is mutation an appropriate tool for testing experiments? Proc. ICSE **2005**, 402–411 (2005)
4. Vincenzi, A.M.R., Simão, A.S., Delamaro, M.E., Maldonado, J.C.: Muta-Pro: towards the definition of a mutation testing process. J. Braz. Comput. Soc. **12**, 49–61 (2006)
5. Granda, M.F., Condori-Fernandez, N., Vos, T.E.J., Pastor, Ó.: Mutation operators for UML Class Diagrams. In: CAiSE 2016 (2016)
6. Object Management Group: Unified Modeling Language (UML) (2015)
7. Object Management Group: Action Language for Foundational UML (ALF) (2013)
8. Object Management Group: Semantics of a Foundational Subset for Executable UML Models (fUML) (2012)
9. Sauro, J.: Estimating productivity: composite operators for Keystroke Level Modeling. In: Human-Computer Interaction. New Trends, pp. 1–10 (2009)
10. Card, S.K., Moran, T.P., Newel, A.: The keystroke-level model for user performance time with interactive systems. Commun. ACM **23**, 396–410 (1980)
11. Haunold, P., Kuhn, W.: A keystroke level analysis of a graphics application: manual map digitizing. In: CHI '94, pp. 337–343 (1994)
12. Teo, L., John, B.E.: Comparisons of keystroke-level model predictions to observed data. In: CHI '06, pp. 1421–1426 (2006)
13. Kieras, D.: Using the keystroke-level model to estimate execution times (2001)
14. Granda, M.F., Condori-fernández, N., Vos, T.E.J., Pastor, O.: What do we know about the Defect Types detected in Conceptual Models ? In: IEEE 9th International Conference on Research Challenges in Information Science (RCIS), pp. 96–107. IEEE, Athens (2015)
15. Tort, A., Olivé, A.: Case Study: Conceptual Modeling of Basic Sudoku. http://guifre.lsi.upc.edu/Sudoku.pdf
16. van Solingen, R., Berghout, E.: The Goal/Question/Metric Method—A Practical Guide for Quality Improvement of Software Development. McGraw-Hill (1999)

17. España, S., González, A., Pastor, Ó., Ruiz, M.: Technical Report Communication Analysis and the OO-Method: Manual Derivation of the Conceptual Model the SuperStationery Co. Lab Demo, Valencia (2011)
18. España, S., González, A., Pastor, Ó., Ruiz, M.: Integration of Communication Analysis and the OO-Method: Rules for the manual derivation of the Conceptual Model, Valencia (2011)
19. Holleis, P., Otto, F., Hussmann, H., Schmidt, A.: Keystroke-level model for advanced mobile phone interaction. In: CHI '07 Proceedings of the SIGCHI Conference on Human factors in Computing Systems 1505–1514 (2007)
20. Granda, M.F., Condori-Fernandez, N., Vos, T.E.J., Pastor, O.: Towards the automated generation of abstract test cases from requirements models. In: 1st International Workshop on Requirements Engineering and Testing. pp. 39–46. IEEE, Karlskrona (2014)

An Open Platform for Studying and Testing Context-Aware Indoor Positioning Algorithms

Nearchos Paspallis and Marios Raspopoulos

1 Introduction

The latest advances in Wireless Communication Systems and in Information Technology gave rise to various applications which require accurate information about the location of the connected devices. Especially in the context of mobile computing and the Internet of Things (IoT) in areas where satellite-based systems fail to provide accurate localization, indoor positioning is regarded as a key enabling technology [1].

The indoor positioning methods and algorithms proposed and developed either experimentally or commercially over the last decades, make use of various kinds of location-dependent radio context such as the Received Signal Strength (RSS), the Time of Arrival (ToA), the Time Difference of Arrival (TDoA), the Angle of Arrival (AoA) etc. [2, 3]. Fingerprint-based positioning [4] has become a very popular topic of research in indoor positioning. It consists of two main phases: the *offline* phase where pre-measured location-dependent information (e.g. RSS), known as fingerprints that cover the entire area of interest, are stored in the database (radio-map), and the *online* phase where the instantaneous measurement is correlated with the fingerprints in the radio-map to estimate the position. These offline and online phases are reminiscent of the training and application phases commonly found in machine learning algorithms.

A prior version of this paper has been published in the ISD2016 Proceedings (http://aisel.aisnet.org/isd2014/proceedings2016).

N. Paspallis (✉) · M. Raspopoulos
University of Central Lancashire, Preston, UK
e-mail: npaspallis@uclan.ac.uk

M. Raspopoulos
e-mail: mraspopoulos@uclan.ac.uk

© Springer International Publishing Switzerland 2017
J. Gołuchowski et al. (eds.), *Complexity in Information Systems Development*,
Lecture Notes in Information Systems and Organisation 22,
DOI 10.1007/978-3-319-52593-8_3

Fingerprint–based positioning using RSS can be classified into two main categories: deterministic and probabilistic. Deterministic methods estimate the location as a convex combination of the reference locations [5]. A very popular technique is the K-Nearest Neighbor (KNN) algorithm which averages the locations of the K fingerprints in the radio-map that better match the received measurement. In the probabilistic approach the location can be estimated by calculating and maximizing the conditional posterior probabilities given an observed fingerprint and a radio-map. This is usually a Bayesian Inference problem in which *a priori* knowledge can be introduced by defining different probabilities to different locations in the environment. Indoor positioning could be device-based in which the device collects the necessary information in order to perform the position estimation on its own or infrastructure-based where the context is pushed to the infrastructure (e.g. a centralized server) which performs the positioning.

In all the above techniques, the sole utilization of the radio parameters, during the position estimation, imposes limits that are hard to overcome. These limitations are often related to the inability of off-the-shelf mobile devices (which are usually based on the IEEE 802.11 standard) to accurately measure these parameters. To this end, the current research trend in indoor positioning is to put forward solutions which enable data fusion of radio-context with non-radio context such as information from inertial sensors and/or any other built-in sensors on the mobile or even context which manually added as prior knowledge (e.g. environment maps) [6].

Various advanced positioning algorithms have been proposed in the literature. These combine various types of information to provide more accurate results. For example, *Inertial Navigation Systems* (INSs) combining different kinds of sensors (e.g. accelerometers, magnetometers, and gyroscopes) have proven to adequately complement existing navigation means such as GPS/GNSS and the same concept can be applied to indoor positioning. For instance, the authors of [7] combine a RSS fingerprinting positioning algorithm with a Kalman filter-based tracking algorithm which estimates the location based on the information collected from inertial sensors.

Generally, the more information that is being considered into a position estimation problem, the higher the probability of a more accurate result. Additionally, the performance evaluation of wireless positioning systems is considered as a challenge by the research community, due to the diversity of positioning algorithms and the complexity of the factors that affect their performance. For this reason, in this paper we propose an open platform which enables easy collection and sharing amongst the research community of context useful for positioning algorithm testing and experimentation.

The rest of this article is organized as follows: Sect. 2 follows up with a description of related work and then Sect. 3 presents the design and implementation of the mobile platform. An evaluation via a custom-tailored experiment is presented in Sect. 4 and the papers closes with conclusions in Sect. 5.

2 Related Work

2.1 Context-Aware Positioning

Context-aware Positioning has attracted significant interest by the research community over the last period The need for this additional context was on one hand aiming to improve the energy efficiency of satellite-based navigation systems and on the other hand to provide additional knowledge into the position estimation process for indoor navigation systems in an attempt to improve the accuracy. An example of the first case is presented in [8] where the duty cycle of energy hungry GPS receivers is reduced by introducing context received from inertial sensors and positioning is performed in a hybrid manner. Information collected from inertial sensors was also proposed and used either in standalone inertial navigation systems or as complimentary context to indoor navigation solutions [9]. Moreover, for indoor navigation, many attempts have been made towards improving the accuracy of fingerprint-based positioning techniques by introducing knowledge extracted from the environment of the device to be positioned [6]. A basic limitation of fingerprint-based techniques lies on the fact that the device heterogeneity may degrade the positioning performance when the device to be positioned is different from the device that was used to collect the radio-map. Differences may arise due to varying antenna characteristics of the mobile terminals which are usually difficult to know or predict. The authors of [10, 11] have proposed the use of linear data transformation to match the characteristics of various devices by collecting only a small set of measurements using the device that is to be positioned, to calibrate a fingerprinting database which has been collected using another device. This effectively means that the type of device is a parameter that needs to be recorded during the data acquisition phase. Also, the orientation of the device is of high importance in fingerprinting positioning and in many cases measurements collected under various orientations at the same location have significant differences and therefore constitute different fingerprints [12].

Other types of context collected from audio, ambient light or other sensors or from any other built-in technology (e.g. Bluetooth) can be used to provide that extra knowledge towards a more accurate position estimation. Basically, any type of additional knowledge which would potentially give an indication about the user whereabouts can be used in conjunction with the positioning estimation process to lead to better results. For example, the authors of [13–15] propose the use of auditable sound to perform or to assist the positioning estimation. The device microphone can be used to identify the rooms that the users are currently located in by matching known sounds (e.g. the sound of the washing machine). With regards to ambient light use in positioning, work in [16] reports that ambient intensity measurements have high location dependency, and they can be used for positioning with the traditional fingerprinting approach. Also, total ambient light irradiance intensity can be used to detect the proximity of a lighting source, and a location can be further resolved with the support of knowledge about the location of the lighting infrastructure. Various attempts were also reported in literature to combine context from heterogeneous radio technologies like Bluetooth, RFID etc. Such a hybrid positioning system [17] achieves better positioning accuracy by exploiting the varying capabilities of the different technologies; that is, Wi-Fi facilitates finger-printing positioning whereas Bluetooth—as a short-range radio technology—allows the partitioning of the indoor space as well as the large Wi-Fi radio-map by using known Bluetooth hotspots. In a similar fashion, the authors of [18] have demon-strated accuracy improvements in the fingerprint–based indoor positioning process, by imposing map-constraints into the positioning algorithms in the form of a–priori probabilities which reflect the probability of a user, to be located on one position instead of all others. These probabilities could be manually set during the off-line phase or they could be dynamically inferred during the on-line phase. An example could be that a professor is more likely to be in his office during his office hours rather than any other place on a University campus.

2.2 Open-Platforms for Positioning

A platform for evaluating positioning on Android devices called *Airplace* was proposed by the authors of [19]. The platform is a mobile-based network-assisted architecture which includes an RSS logger, a radio-map distribution server and a "Find Me" application which facilitates the testing of various positioning algo-rithms and the optimization of their settings. The authors of [20] have presented *Anyplace* [21] a free and open navigation service that relies on the abundance of sensory data on smartphones to deliver reliable indoor geolocation information. It implements a set of crowdsourcing-supportive mechanisms to handle the enormous amount of crowd-sensed data. A similar open platform is presented in [22]. This is called SmartCampusAAU and it facilitates the creation of indoor positioning sys-tems. It includes an application and a back-end that can be used to enable device- or infrastructure-based indoor positioning and a publicly available Open Data backend to allow researchers to share radio map and location tracking data. The platform relies on crowdsourcing techniques to construct radio-maps. Crowdsourcing [23, 24] leverages the positioning fingerprints collected by users using their smart-devices in order to construct and/or update the radio-map. This obviously presents various inaccuracies that need to be considered in the position estimation phase, mainly the fact that the radio-map will not be homogeneous as it contains fingerprints from a diverse set of devices. The authors of [25] have tackled this problem by collecting signal differences instead of absolute signal strength values. In the crowdsourcing process, recording the type and model of the devices is of particular importance.

In our approach we take this open-platform concept one step further by intro-ducing additional non-radio location-dependent context which can be openly used for developing advanced positioning algorithms, fusing together various kinds of information towards a more accurate position estimate.

3   Platform Design and Implementation

3.1   Administration

The platform is designed as a mobile-based system which can collect, store and process data autonomously while offline. The users define their own named locations (e.g. typically a location corresponds to a building, or a group of neighboring buildings such as a campus). Each location must feature at least one floor, but possibly more (see Fig. 1).

The users are asked to provide their blueprints for each floor/level, and specify the exact coordinates for the upper left and the lower right corners of the image (see right screenshot in Fig. 1 and left screenshot in Fig. 2). With this information and assuming that the blueprint image is north-aligned, the system can then associate each point on the image with the corresponding geographic coordinates (i.e. latitude and longitude). This is important as it allows the users to easily specify their actual position during the training of the system, using a visual targeting system (see right screenshot in Fig. 2).

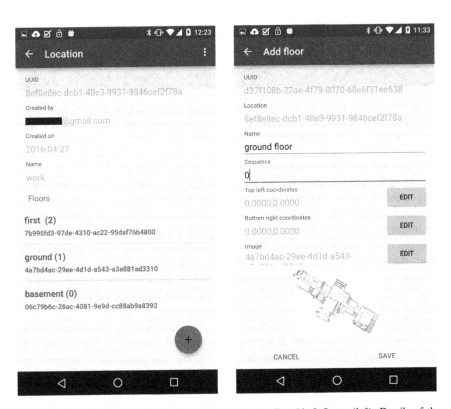

Fig. 1 A screenshot of a view of a location and its details with 3 floors (*left*). Details of the selected 'ground' floor with its coordinates—still undefined—and a blueprint (*right*)

Fig. 2 Aligning the blueprint with the real world using satellite imagery (*left*). In the training phase, the users select their exact position—at the correct floor—and then start a new scan. Optionally, the user can select an automatically triggered scan that repeats periodically (*right*)

In the training (or offline) phase, the user collects fingerprints with the aim of generating the training data for the algorithm. Each fingerprint is associated with a well-defined location on the indoor map and more specifically on a selected floor of the given location. The user utilizes a crosshair-like target and a draggable view of the blueprint (usually a floor map) to specify the exact position of the user at the time the fingerprint is collected. While the crosshair is fixed at the center of the view, the user is able to drag the underlying view to select her or his current position in the building (see right screenshot in Fig. 2). It must be assumed that the users doing the training have good knowledge of the building and are reasonably able to navigate inside the building using the floorplan.

While a fingerprint typically includes only the signal strength from nearby Wi-Fi access points, the user can also specify additional context to be stored. The available context data include information that has the potential to improve the accuracy of a positioning algorithm, such as the make and model of the device, environmental data such as temperature, pressure, humidity, inertial data such as accelerometer and gyroscope readings, etc. The selection of which data to collect, is configurable via a settings screen (see left screenshot in Fig. 3).

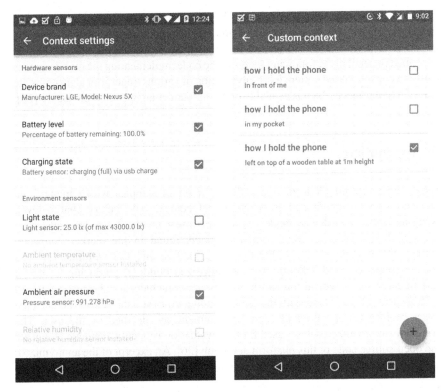

Fig. 3 The context settings screen which allows users to preview and select the exact subset of context data they want included in the fingerprint—only sensors supported by the device's hardware are enabled (*left*). A custom context screen allows the users to define extra tags and use them to annotate the collected fingerprints with context otherwise hard to infer (*right*)

In some cases, users would want to annotate their collected fingerprints with additional information which is otherwise hard to automatically infer using the device's sensors. For example, a user might want to indicate whether they are holding the device at waist or eye height, hypothesizing that this might affect the received signal strength. For this reason, the app features a mechanism where users can define their own custom tags, and then select which ones to include in their training session, as needed (see right screenshot of Fig. 3). The users can manually select which annotations to be included during training by adding their tags, and selecting the corresponding checkboxes.

In the application (or online) phase, the user can use the system to determine her or his position in real time. Typical positioning algorithms achieve this by collecting a fingerprint from their present location, and comparing it to those stored during the training phase. The closest match (or the average of the K closest matches in the case of the K-Nearest Neighbor algorithm) is then shown as the present location on the corresponding fingerprint. A few standard algorithms are evaluated in Sect. 4, but the system is designed to easily accommodate additional, custom algorithms.

While in its final form, the platform is envisioned to allow the use of several interchangeable algorithms selectable in runtime, in its current form the data must first be exported into a JSON-formatted file, and then be processed offline. This is described in detail in the following section. The code implementing the mobile app, as well as a console-based system for the evaluation of various fingerprint algorithms are part of a live project which is available on Github under an open-source license [26].

4   Evaluation

Using the proposed platform, we collected a set of samples which include the received signal strength and some additional context (i.e. battery charge level, charging status, and device model). The experiment involved 4 individuals, each using a different device from common models found in the market: Samsung Galaxy S2, Google Nexus 4, Prestigio Multipad 8, and LG Optimus L5. The measurements covered 2 floors of a medium-sized building, producing a total of 307 fingerprints. A typical fingerprint measurement includes a radio-map of 10–20 RSS measurements, along with the corresponding context data. It also includes the exact coordinates and floor of each measurement, as specified by the user. The complete, anonymized dataset used for this evaluation is openly available alongside the open-source code of this platform at Github [26]. An excerpt of the anonymized, JSON-formatted file is listed below for reference:

```
{
  "trainings": [
    {
      "uuid": "agtzf...PKtgwkM",
      "floorUUID": "a",
      "createdBy": "anonymized",
      "timestamp": 1403776198856,
      "lat": 35.008370626294614,
      "lng": 33.69661813690848,
      "context": { "batteryCharging": "false",
          "batteryLevel": "27.0", "model": "PMP5880D",
          "product": "asbis" },
      "measurements": [
          {"macAddress": "0", "ssid": -65 },
          {"macAddress": "1", "ssid": -66 },
  ...

          {"macAddress": "19", "ssid": -82 }
      ]
    },
    ...
  ]
}
```

Table 1 Comparing the performance of a few fingerprinting algorithms

	Standard deviation		Absolute values	
	Mean	Variance	Min	Max
Standard fingerprint algorithm	8.7004	28.3425	1.1740	22.9706
Same device only	10.1144	78.1406	2.4748	40.0940
Similar battery only (±10%)	10.0622	66.9148	2.3564	37.4445
K-Nearest Neighbor with K = 3	7.6623	22.6163	0.9552	21.3290
K-Nearest Neighbor with K = 10	7.8376	21.8472	1.1825	21.0465

The measurements show the distance in meters from the actual target (i.e. the error) of each algorithm for a randomly chosen subset (of size 10%) of the fingerprints. All distances are in meters

To assess the quality of the dataset and the effectiveness of various fingerprinting algorithms, we implemented the following experiment: First, the collected dataset of 307 fingerprints was randomly split to two subsets: 90% for training and 10% evaluation/application. Then, each of the application fingerprints was compared to the training fingerprints, and the location was determined according to the logic of the used algorithm. The resulting error of each algorithm (i.e. distance of predicted versus actual position, in meters) is summarized in Table 1.

In its simplest form, the *Standard fingerprint algorithm* uses the Euclidean distance to measure the distance among individual readings in the radio-map, as described in [27]. When comparing fingerprints, missing access point readings were assumed to have a zero RSS. The *Same device only* variant uses the exact same mechanism, but filters training fingerprints to select only those that were generated by the same device model. Similarly, the *Similar battery only* variant filters fingerprints to select those only that were generated by devices which (at the time of the training) had similarly charged batteries. The last two variants, refer to using the K-Nearest Neighbor algorithm with K equals 3 and 10 respectively. In this case, the algorithm first identifies the best K matches (i.e. best 3 or 10 accordingly) using the standard fingerprint algorithm, and then decides the inferred position to be the center of those best matches.

Using the samples collected in our experiment, we applied each of the algorithms described above, and measured their perceived accuracy. Table 1 lists the performance of each of the algorithms in terms of the mean distance and the variance for all tests, using standard deviation calculations on the resulting measurements of the distance error. For reference, the table also lists the minimum and the maximum distances measured. All measurements in Table 1 are in meters.

Studying these measurements reveals that the standard fingerprint algorithm works quite effectively and precisely already in its simplest form. Utilizing additional context information such as the device model or the battery charge level did not appear to improve the accuracy of the algorithm, which can be partly explained by the fact that the training phase was rather dense (with multiple fingerprints in each room, for each user and each device model). This is also evident in the fact that the K-Nearest Neighbor-based techniques have performed better than all others (especially the *best of 3* variant), indicating that the fingerprints are densely covering the building's area.

The experiment has shown that the presented platform provides a reliable and convenient mechanism for collecting fingerprints that are accurate and include several context data. We argue that this will assist in creating an open ecosystem where multiple shared datasets can be utilized to evaluate new fingerprint-based algorithms, and/or fine-tune existing ones.

5 Conclusions

Indoor positioning is an increasingly important topic, with both a social and financial impact. Most commonly, fingerprint-based algorithms are used to enable indoor positioning, where the fingerprints consist primarily of radio-maps. Nevertheless, with the advancement of sensory technology on new smart-phone devices, algorithms are increasingly dependent on additional context information to increase their accuracy. Even though many such algorithms have been developed, very few have been tested or were applied to real-world conditions.

In this paper we present an open platform which enables researchers and practitioners to easily perform their own experiments using fingerprint datasets which fuse together radio-specific context with additional context that could potential offer better positioning estimates. Users are enabled to quickly set up their locations, floors etc. and then perform the training. The collected data can be easily exported as JSON-formatted data which can then be used for offline assessment of the accuracy of arbitrary algorithms. Additionally, generated datasets can be shared with the scientific community. In a preliminary evaluation, we have shown how the collection of such data can enable testing arbitrary fingerprinting algorithms, as well as variants of them, in a straightforward and convenient way.

In this ongoing project, we envision to further enhance the mobile platform and integrate it with a cloud-based system for storing, sharing and making datasets openly available. We also aim at enabling the collection of additional type of context, as well as covering more extensive time periods and more device models, while in the process producing exemplar open research datasets.

References

1. Macagnano, D., Destino, G., Abreu, G.: Indoor positioning: a key enabling technology for IoT applications. In: IEEE World Forum on Internet of Things (WF-IoT), Seoul (2014)
2. Jami, I., Ali, N.M.F.N.M., Ormondroyd, R.F.: Comparison of methods of locating and tracking cellular mobiles. In: IEE Colloquium on Novel Methods of Location and Tracking of Cellular Mobiles and Their System Applications (Ref. No. 1999/046), London (1999)
3. Rappaport, T.S., Reed, J.H., Woerner, B.D.: Position location using wireless communications on highways of the future. IEEE Commun. Mag. **34**(10), 33–41 (1996)

4. Bahl, P., Padmanabhan, V.N.: RADAR: an in-building RF-based user location and tracking system. In: 19th Annual Joint Conference of the IEEE Computer and Communications Societies, Tel Aviv (2000)
5. Honkavirta, V., Perala, T., Ali-Loytty, S., Piche, R.: A comparative survey of WLAN location fingerprinting methods. In: 6th Workshop on Positioning, Navigation and Communication (WPNC), Hannover (2009)
6. Raspopoulos, M., Denis, B., Laaraiedh, M., Dominguez, J., De Celis, L., Slock, D., Agapiou G., Stephan, J.S.S.: Location-dependent information extraction for positioning. In: International Conference on Localization and GNSS, Starnberg (2012)
7. Sangwoo, L., Bongkwan, C., Bonhyun, K., Sanghwan, R., Jaehoon, C., Sunwoo, K.: Kalman filter-based indoor position tracking with self-calibration for RSS variation mitigation. Int. J. Distrib. Sens. Netw. **11**(8), 180 (2015)
8. Jurdak, R., Corke, P., Dharman D., Salagnac, G.: Adaptive GPS duty cycling and radio ranging for energy-efficient localization. In: The 8th ACM Conference on Embedded Networked Sensor Systems (SenSys 2010), Zurich, Switzerland (2010)
9. Harle, R.: A survey of indoor inertial positioning systems for pedestrians. IEEE Commun. Surv. Tutorials **15**(3), 1281–1293 (2013)
10. Raspopoulos, M., Laoudias, C., Kanaris, L., Kokkinis, A., Panayiotou C.G., Stavrou, S.: 3D Ray Tracing for device-independent fingerprint-based positioning in WLANs. In: 9th Workshop on Positioning Navigation and Communication (WPNC), Dresden (2012)
11. Raspopoulos, M., Laoudias, C., Kanaris, L., Kokkinis, A., Panayiotou C.G., Stavrou, S.: Cross device fingerprint-based positioning using 3D Ray Tracing. In: 8th International Wireless Communications and Mobile Computing Conference (IWCMC), Limassol (2012)
12. Su, D., Situ Z., Ho, I.W.-H.: Mitigating the antenna orientation effect on indoor Wi-Fi positioning of mobile phones. In: IEEE 26th Annual International Symposium on Personal, Indoor, and Mobile Radio Communications (PIMRC), Hong Kong (2015)
13. Mandal, A., Lopes, C.V., Givargis, T., Haghighat, A., Jurdak, R., Baldi, P.: Beep: 3D indoor positioning using audible sound. In: Second IEEE Consumer Communications and Networking Conference CCNC, Las Vegas (2005)
14. Madhavapeddy, A., Scott D., Sharp, R.: Context-aware computing with sound. In: 5th International Conference on Ubiquitous Computing, Seattle (2003)
15. Joe, C., Yip, L., Elson, J., Wang, H., Maniezzo, D., Hudson, R.E., Yao, K., Estrin, D.: Coherent acoustic array processing and localization on wireless sensor networks. Cent. Embed. Netw. Sens. **91**(8), 1154–1162 (2003)
16. Liu, J., Chen, Y., Jaakkola, A., Hakala, T., Hyyppa, J., Chen, L., Chen, R., Tang, J., Hyyppa, H.: The uses of ambient light for ubiquitous positioning. In: IEEE/ION Position, Location and Navigation Symposium (PLANS2014), Monterey (2014)
17. Baniukevic, A., Jensen, C.S., Lu, H.: Hybrid indoor positioning with wi-fi and bluetooth: architecture and performance. In: IEEE 14th International Conference on Mobile Data Management (MDM), Washington, DC (2013)
18. Kokkinis, A., Raspopoulos, M., Kanaris, L., Liotta, A., Stavrou, S.: Map-aided fingerprint-based indoor positioning. In: IEEE 24th International Symposium on Personal Indoor and Mobile Radio Communications (PIMRC), London (2013)
19. Laoudias, C., Constantinou, G., Constantinides, M., Zeinalipour-Yazti, N.S.D., Panayiotou, C.: The airplace indoor positioning platform for android smartphones. In: 2012 IEEE 13th International Conference on Mobile Data Management, Bengaluru, Karnataka (2012)
20. Demonstration Abstract: Crowdsourced Indoor Localization and Navigation with Anyplace International conference on Information processing in sensor networks. In: IPSN'14, IEEE Press, Berlin, Germany (2014)
21. [Online]. Available: https://anyplace.cs.ucy.ac.cy/
22. Hansen, R., Thomsen, B., Thomsen, L.L., Stubkjær, F.: SmartCampusAAU—an open platform enabling indoor positioning and navigation. In: 14th International Conference on Mobile Data Management, Milan (2013)

23. Mazumdar, P., Ribeiro V.J., Tewari, S.: Generating indoor maps by crowdsourcing positioning data from smartphones. In: International Conference on Indoor Positioning and Indoor Navigation (IPIN), Busan (2014)
24. Wu, C., Yang, Z., Liu, Y.: Smartphones based crowdsourcing for indoor localization. IEEE Trans. Mob. Comput. **14**(2), 444–457 (2015)
25. Laoudias, C., Zeinalipour-Yazti, D., Panayiotou, C.G.: Crowdsourced indoor localization for diverse devices through radiomap fusion. In: International Conference on Indoor Positioning and Indoor Navigation, Montbeliard-Belfort (2013)
26. Paspallis, N.: Context-aware indoor positioning system. [Online]. Available: https://github. com/nearchos/CAIPS. Accessed 03 Oct 2016
27. Varshavsky, A., Patel, S.: Location in ubiquitous computing. In: Krumm, J. (ed.) Ubiquitous Computing Fundamentals, pp. 285–320. Chapman and Hall/CRC, Boca Raton (2009)

Automation of the Incremental Integration of Microservices Architectures

Miguel Zúñiga-Prieto, Emilio Insfran, Silvia Abrahão and Carlos Cano-Genoves

1 Introduction

The need to maintain high customer satisfaction by delivering new or customized products and services signifies that development paradigms are changing to the Continuous Integration (CI) and Continuous Deployment (CD) of software functionality, in which companies offering internet-based services should be capable of providing customers with software functionality on a daily basis [1]. The microservice architectural style has therefore emerged to facilitate CI/CD by affecting the way in which software development teams are structured, source code is organized and continuously built/packed, and software products are continuously deployed [2]. This architectural style proposes the development of a single application as a suite of small and cohesive sets of microservices built around business

A prior version of this paper has been published in the ISD2016 Proceedings (http://aisel.aisnet.org/isd2014/proceedings2016).

M. Zúñiga-Prieto (✉)
Department of Computer Science, Universidad de Cuenca, Cuenca, Ecuador
e-mail: miguel.zunigap@ucuenca.edu.ec

E. Insfran (✉) · S. Abrahão (✉) · C. Cano-Genoves
Department of Information Systems and Computation, Universitat Politècnica de València, Valencia, Spain
e-mail: einsfran@disc.upv.es

S. Abrahão
e-mail: sabrahao@disc.upv.es

C. Cano-Genoves
e-mail: carcage1@inf.upv.es

© Springer International Publishing Switzerland 2017
J. Gołuchowski et al. (eds.), *Complexity in Information Systems Development*,
Lecture Notes in Information Systems and Organisation 22,
DOI 10.1007/978-3-319-52593-8_4

51

capabilities, and independently developed, deployed and scaled, thus allowing them to scale their applications, gain agility and get new functionalities out to customers faster [3, 4].

The flexibility in resource management (e.g. processing, memory, message queues) provided by cloud environments is motivating organizations to consider them as their systems deployment environment, in which different Infrastructure as a Service (IaaS) or Platform as a Service (PaaS) environments are chosen depending on Service Level Agreements (SLA) or other requirements. Cloud environments are a well suited option as regards deploying microservices [5, 6], since they allow companies to gain agility and reduce complexity not only when deploying and scaling microservices, but also by acquiring resources provisioned according to specific microservice needs. However, applications that will be deployed in cloud environments (*cloud applications*) must be developed using cloud vendor standards, thus preventing developers from creating software that can be deployed on multiple clouds, which is known as *vendor lock-in* [7]. The incremental nature of the microservice-based applications development additionally leads to a situation in which the application's architecture evolves each time a microservice is integrated into it. Building microservices for deployment in cloud environments therefore requires managing architectural changes (architectural reconfiguration) and minimizing application disruptions while the integration takes place.

Current cloud development approaches do not support microservice development/migration and only a few technical reports on this can be found (e.g., [6, 8, 9]). Approaches that support the development of cloud applications are related to this work (e.g., [10–12]); however, proposals confronting the incremental development and its architectural implications are still lacking. Furthermore, in terms of architectural reconfiguration, as far as we know, there are no proposals that support a systematic reasoning about the architectural impact of the integration of the services included in a given software increment into the current application architecture. In previous works [13, 14], we introduced a general process definition for the DIARy method which follows an incremental and model driven development approach that supports the incremental integration of cloud service applications and their dynamic architecture reconfiguration triggered by the integration of new *software increments* (increments); to support the specification and generation of some software artifacts for service architecture reconfiguration. In this paper, we extend the DIARy process by defining new activities and tasks to satisfy microservices principles and support the incremental integration of microservices, covering the lack of proposals that allow developers to propagate integration design decisions to reliable software artifacts that improve the agility of integration and deployment processes. We also provide the tool support needed to automate these tasks by defining models that describe the microservice integration logic, as well as the transformation chains, which automate the generation of software artifacts that implement the integration logic (orchestration among microservices), and scripts for architectural reconfiguration.

The remainder of this paper is structured as follows. Section 2 contains a description of the background and discusses related works. Section 3 presents an overview of the method proposed. Section 4 illustrates the use of our approach in a case study. Finally, Sect. 5 presents our conclusions and future work.

2 Background and Related Work

The microservice architectural style is a lightweight subset of the Service Oriented Architecture (SOA), in which: "the main difference between SOA and microservices is that the latter should be self-sufficient and deployable independently of each other while SOA tends to be implemented as a monolith" [15]. This architectural style is gaining acceptance as regards overcoming the shortcomings of a monolithic architecture in which, rather than having the application logic within one deployable unit, applications are decomposed into services, each of which is deployable on a different platform, runs its own process, and communicate by means of lightweight mechanisms. The main principles of microservices are [3]:

- *Componentization* via *Services*: Software is broken up into multiple services that are independently replaceable and upgradeable and communicate by means of inter-process communication facilities using an explicit component-published-interface.
- *Organized Around Business Capabilities*: Microservices are implemented around business areas, in which services include a user-interface, storage, and any external collaborations.
- *Products not Projects*: Development teams own a product throughout its entire lifetime, taking full responsibility for the software in production.
- *Smart Endpoints and Dumb Pipes*: Business logic, related business rules, and data reside in the services themselves rather than in a centralized middleware. Simple messaging or a lightweight messaging bus is used to provide communication among microservices.
- *Decentralized Governance*: Standardization on a single technology platform is avoided; the right technological stack for a job should be used, and each microservice manages its own decisions regarding tools, languages, and data storage.
- *Decentralized Data Management*: Decisions concerning both the conceptual model of the world and data storage will differ between microservices.
- *Infrastructure Automation*: Automatic means to integrate and deploy in new environments.
- *Evolutionary Design*: Services are independently replaced and upgraded, which is achieved by using service decomposition as a tool so as to enable application developers to control changes in software applications at the pace of business changes.

Decentralized Governance and *Decentralized Data Management* microservice principles suggest avoiding standardization in a single technology; however, certain development challenges (e.g., the *vendor lock-in*) need to be addressed in order to produce services that are feasible for deployment in different cloud environments. Furthermore, the *Infrastructure Automation* microservice principle suggests having an automatic means of integration and deployment in new environments. However, despite the fact that development teams building microservices use CI/CD techniques and tools [3], these techniques require the inclusion of reliable software artifacts (e.g., implementation code, deployment scripts, configuration scripts) in their automated building processes or deployment pipelines. Software artifacts should therefore be error free in order to ensure that the CI/CD's automated test functionalities do not prevent the integration or deployment process. Finally, CI/CD requires the making of architectural decisions [15], where in a context in which the application architecture evolves with each microservice integration, mechanisms that support the specification of architectural decisions and manage architectural changes without preventing the execution of applications are therefore required.

Model-Driven Development (MDD) is an approach used to develop software systems in which developers build an application by refining models at different levels of abstractions, and then obtain implementation artifacts by means of model transformations. We believe that an MDD approach provides good support as regards managing microservice integration and the consequent architectural evolution of the application. This approach will allows developers to: (i) capture technology-independent microservice integration specification and deployment information, thus making design artifacts reusable and enabling developers to overcome the *vendor lock-in* issue; (ii) propagate microservice integration specification to implementation/deployment/reconfiguration artifacts, thus enabling developers to obtain error free artifacts; and (iii) automate building, packaging, deployment and the architectural reconfiguration process.

2.1 Related Work

Developing applications by using the microservice architectural style is a relatively new approach, and only a few related technical reports can be found (e.g., [6, 8, 9]). These works describe design decisions made or strategies employed in order to either satisfy microservice principles, or make use of CI/CD tools and techniques; however, they do not propose design, implementation or integration methods. Moreover, those works do not propose mechanisms with which to help to obtain error free artifacts to be included into CI/CD pipelines.

Microservices are cloud-native architectures, and the MDD approaches that support the development of cloud applications are therefore related to this work (e.g., [10–12, 16]). These approaches apply MDD principles in order to tackle the *vendor lock-in* problem when developing or migrating cloud applications. With regard to approaches that propose mechanisms with which to document design decisions in cloud environments we can highlight CAML [17], MULTICLAPP [18] and CloudML [19]. These works define UML profiles or other modeling languages used to describe deployment topologies, applications as a composition of software artifacts to be deployed across multiple clouds, or resources that a given application may require from existing clouds. However, although "getting integration right is the single most important aspect of the technology associated with microservices" [4], these proposals do not provide mechanisms with which to specify architectural decisions regarding integration and the impact of integrating increments in the current cloud application architecture. Finally, with regard to approaches for dynamic reconfiguration, works such as SeaClouds [20] or MODAClouds [12] propose mechanisms that can be used to achieve architectural reconfiguration either by replacing orchestration or as result of the re-deployment of components. These proposals do not allow the specification of the architectural changes produced during integration nor do they take into account implementation alternatives that facilitate scalability and the re-deployment of services in different clouds.

3 A Method for the Incremental Integration of Microservices

This method allows cloud applications to be constructed as a composition of microservices, in which each microservice design is included in an incremental increment integration process that allows architects to specify how microservices will be integrated into a cloud application. Developers use the increment integration specification to generate software artifacts, such as skeletons of microservice logic, interaction protocol and scripts with which to build, deploy and architecturally reconfigure the current cloud application, all of which are generated according to each microservice technology specification. In order to define this method, we analyzed how our previous work satisfies the principles of the microservice architectural style; we then used the lessons learned to extend the DIARy-process [13, 14]. The Microservice Incremental Integration Method, which is made up of the Microservices Incremental Integration Process (also referred to as the *Integration Process*), the adapted DIARy-specification-profile [21] and transformation chains, is explained as follows. Figure 1 shows the *Integration Process*, whose main activities are explained in the next sections.

3.1 Increment Integration Specification

This activity aims to allow architects to specify how to integrate a microservice (*Microservice Architecture Model*) into the current application (*Application Architecture Model*) by specifying both the integration logic and the architectural

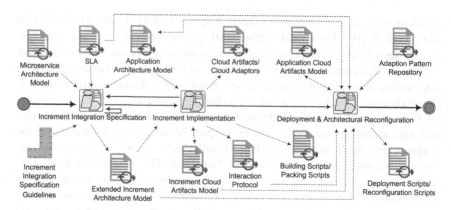

Fig. 1 The Microservice Incremental Integration Process

impact of integration, without taking into consideration the specifics of any cloud environment. This is an iterative activity that provides architects with the possibility of specifying the integration of increments composed of several microservices; therefore architects take *Microservice Architecture Models* as input, include them as part of an increment's architecture (*Extended Increment Architecture Model*), follow *Microservices Composition and Increment Integration Specification Guidelines*, and make integration design decisions based on SLA terms (whose definition, specification and representation is outside of this work scope). The DIARy-Specification-profile (see [21] for more details about its usage) helps architects to create the *Extended Increment Architecture Model* which specifies the increment integration by documenting the increment's architecture, the integration's logic and the architectural impact of integration. This model complies with *the Extended Increment Architecture Model* metamodel, which is explained below.

The *Extended Increment Architecture Model* metamodel
The Service oriented architecture Modeling Language (SoaML) [22] is an OMG specification that was specifically designed for the modeling of service-oriented architectures. SoaML leverages the Model Driven Architecture (MDA) approach and provides a UML profile and a metamodel that extends the UML metamodel. The DIARy-specification-profile extends the SoaML profile, resulting in an ADL that facilitates the increment integration specification. In order to facilitate software artifact generation, this work extends SoaML and UML metamodels in order to define the *Extended Increment Architecture Model* metamodel (see Fig. 2). Owing to space limitations, Fig. 2 includes only those meta-classes that define the main concepts used to describe integration logic and architectural impact, in which meta-classes belonging to the UML metamodel are depicted with an icon next to the

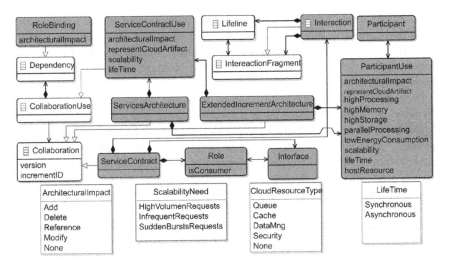

Fig. 2 Excerpt of *Increment Architecture Model* meta-model

meta-class name, whereas meta-classes that extend the SoaML/UML notations are depicted with a background color.

An *Extended Increment Architecture* extends a UML *Collaboration*, thus allowing the increment integration specification by modeling both the integration logic and the architectural impact of integration. Integration logic is described by its inner parts (i.e., *ParticipantUse*, *RoleBinding*, and *ServiceContractUse*) which model the interoperation between *Participants* belonging to the increment with *Participants* belonging to the current *Application Architecture Model*. Additionally, architectural impact is described by tagging its inner parts with *architecturalImpact* values (*Add*, *Modify*, and *Delete*) that describe the architectural change that each inner part will produce on the current *Application Architecture Model* after integration.

A *Participant* represents: (i) a microservice to be integrated, (ii) a microservice/component already existing in the current application architecture with which the microservice(s) to be integrated will interoperate, (iii) a microservice/component to be created in order to consume microservice services or provide it with services, and (iv) a cloud resource consumed by a microservice.

A *ServiceContract* extends a UML *Collaboration* and, as SoaML proposes, represents an agreement between the involved *Participants* about how the service is supposed to be provided and consumed (interoperation). A *Service Contract* definition includes the following inner parts: (i) *Roles* that *Participants* involved in a service must fulfil in order to interoperate, (ii) provided and required *Interfaces* that explicitly model the provided and required operations to complete the service functionality and that *Participants* must implement in order to fulfil a *Role*, (iii) and an *Interaction Protocol* that specifies the interoperation between *Participants* without defining their internal processes. This work uses *Service Contracts* in order to describe integration logic, where an UML activity diagram is used to model the interaction protocol among *Participants* belonging to the increment with *Participants* belonging to the current *Application Architecture Model*. *Service Contracts* are implemented as services that manage interoperation among *Participants*.

Participants and *ServiceContracts* may be reused, therefore a *ParticipantUse* references a *Participant* involved in a specific service and a *ServiceContractUse* explicitly specifies the use of the interoperation described in a *ServiceContract*. The attributes *scalability* and *lifetime* make it possible for architects to specify requirements related to the expected demand of a *Participant* or *ServiceContract* service. The attributes *HighProcessing*, *HighMemory*, *HighStorage*, *Parallel Processing*, and *LowEnergyConsumption* are used to specify characteristics of the cloud resources that a *Participant* is expected to consume from the cloud environment. Finally, the attribute *hostResource* describes the cloud resource type of a *Participant* representing a cloud resource.

The *Services Architecture of the participant*, modeled as a SoaML Services Architecture diagram, specifies how parts of a microservice work together to play the owning microservice's role(s). It includes the microservice's architectural elements as well as interoperation requirements described by *outside Roles* that external *Participants* must play in order to interact with the microservice and *outside ServiceContracts* that describe the interaction among those roles.

Finally, a *RoleBinding* binds each of the *Roles* defined in a *ServiceContract* to a *Participant*, both of which are referenced in an *ExtendedIncrementArchitecture*.

Integration Specification

Integration specification is done by using high level representations of microservices, which is achieved by creating *Participants* that represent a *Microservice Architecture Models*. Consequently, each *Microservice Architecture Model* taken as input in this activity becomes the *Services Architecture of the participant* that represent the microservice to be integrated. Once a *Participant* representing a microservice to be integrated has been created, architects specify the integration logic by creating an *ExtendedIncrementArchitecture* element, and then create its inner parts: (i) a *ParticipantUse* that references a *Participant* representing a microservice to be integrated; (ii) *ParticipantUses* that reference *Participants* belongings to the current *Application Architecture Model* that, by playing *outside Roles*, will interoperate with *Participants* representing the microservice to be integrated; (iii) *ServiceContractUses* that use the interoperation defined in *outside ServiceContracts*; (iv) *RoleBindings* that bind each of the *outside Roles* defined in an *outside ServiceContract* to the *ParticipantUse* that will play the role.

Developers specify the architectural impact of integration by tagging *ExtendedIncrementArchitecture* inner parts with *architecturalImpact* values that describe how they collaborate to reconfigure the current *Application Architecture Model* (e.g., by adding *RoleBindings*, adding *Participants*, removing *Participants*). Finally, for each *ParticipantUse* and *ServiceContractUse*, architects specify theirs expected demand and usage of cloud resources.

In this activity, the creation of *Extended Increment Architecture Models* allows architects to satisfy *Componentization* via *Services* and *Organized around Business Capabilities* microservice principles. Furthermore, designing microservice integration in advance not only facilitates incremental integration of microservices but also allows different development teams working independently on different microservices to take full responsibility for the software in production, satisfying *Evolutionary Design* and *Products not Projects* microservice principles.

3.2 Increment Implementation

This activity aims to support the integration process by generating platform-specific cloud artifacts (software artifacts to be deployed on a cloud platform), includes the following steps:

Check Increment Compatibility

Architects participate in verifying whether the *ExtendedIncrementArchitecture Model* is compatible with the current *ApplicationArchitectureModel*. If discrepancies exist between the *Participant's* interfaces (e.g., different names for methods and services, different message ordering), they design a *ServiceContract* that overrides *outside ServiceContracts* and apply model-to-text (M2T) transformations that generate skeletons of *Cloud Adaptors* (see Fig. 1).

Specify the Packaging and Deployment Structure

In this step, developers apply model-to-model (M2M) transformations to translate the *Extended Increment Architecture Model* into a model that describes the cloud artifacts needed to implement its inner parts (*DIARyArchitecturalElements*): the Increment *Cloud Artifacts Model*. This model organizes cloud artifacts into projects that can be packed/built/deployed independently in different cloud environments in accordance with decisions made during the development process (e.g. technology, microservice workload management decisions). This model promotes the decoupling of software artifacts that implement interaction protocol from those that implement microservice design, thus satisfying the *Smart Endpoints* and *Dumb Pipes* microservice principles. The *Increment Cloud Artifacts Model* complies with the *Cloud Artifacts Model* meta-model (see Fig. 3).

The Cloud Artifacts Model meta-model

The way in which microservices are deployed has an influence on satisfying SLA terms or other nonfunctional requirements [23] (e.g., agility to deploy, modifiability, monitoring, cost of provisioning). We use Projects to manage the building, packaging and deployment options. M2M transformation rules map *Interaction Projects* onto *Service Contracts* (see Fig. 2) architectural elements, and generate descriptions of cloud artifacts that allow developers to implement interoperation among microservices as a separate service. An Interaction Project includes the following cloud artifacts: *Interaction Service—Hosted Services* that implement

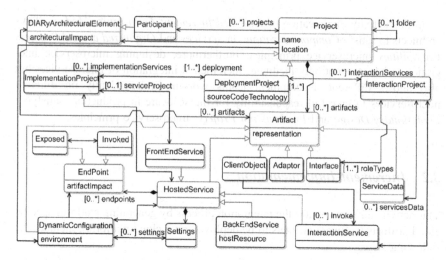

Fig. 3 Excerpt of *Cloud Artifacts Model* meta-model

interoperation interaction protocols, *Interface* definitions, and *Service Data* (message Types or data Types). M2M transformation rules map inner parts of *Service Contracts* architectural elements onto the before mentioned cloud artifacts.

In the case of *Participants* that provide services, *ImplementationProjects* are mapped onto *Service Architectures of the participant* architectural elements (see Fig. 2). These *Implementation Projects* include descriptions of *Artifacts*, such as *FrontEndService—Hosted Services* that implement microservice business logic, *Interface* implementations that implement interfaces defined in related *Service Contracts*, *BackEndService—HostedServices* that use cloud resources (e.g., message queues), or *Adaptors* that correct incompatibilities between interfaces. Additionally, in the case of *Participants* that play a *Role* whose attribute *isConsumer* = true (see Fig. 2), *Implementation Projects* also include *ClientObjectArtifacts* which implement corresponding *Interfaces* and initiate the service execution by invoking *InteractionServices* that manage interoperation (orchestration/choreography). For detailed mappings see [13].

Deployment Projects and *Interaction/Implementation Projects* facilitate the packaging of *Artifacts* into a deployable package. Microservices whose related projects are included into a *Deployment Project* will be implemented with the same technology and deployed in the same cloud environment, whereas *Artifacts* included in an *Interaction/Implementation* project will be packed together in the same deployment artifact and deployed in the same cloud environment resource (e.g., virtual machine), thus sharing cloud environment resources. Including microservice related *Artifacts* in an exclusive or shared *Interaction/Implementation Project* allows developers to manage workload changes and running costs.

DynamicConfiguration meta-classes describe *Settings of HostedServices* (e.g., service parameters) that could change at runtime. *Settings* and *Invoked/Exposed EndPoints* information will therefore be stored outside the deployable package.

Thus enabling them to be updated without requiring the redeployment of the entire package, a best practice in the CD [24].

In order to specify the packaging and deployment structure, we provide an Eclipse plug-in which executes M2M transformations carried out using the Atlas Transformation Language (ATL) to generate Increment Cloud Artifacts Models from Extended Increment Architecture Models. Input/Output models are implemented as ecore models in the Eclipse Modeling Framework (EMF). Transformations generate descriptions of the cloud *Artifacts* required to implement architectural elements and define the packaging/deployment structure by assigning *Artifacts* to different *Interaction/Implementation/Deployment Projects* according to the *architecturalImpact*, and expected demand and usage of cloud resources (e.g., values *scalability*, *lifeTime*, *HighProcessing*). Figure 4 shows an excerpt of the transformation rule applied to assign the *Artifacts* corresponding to *Participants* that require *scalability* = *HighVolumenRequests* (line 4) into an exclusive *ImplementationProject* (line 6) that is assigned to a *DeploymentProject* (line 9) that will be deployed in an exclusive virtual machine. Additionally, in the case of *Participants* that play a *Role* whose attribute *isConsumer* = true, a client object that initiate the interaction is created (line 12).

Generate Implementation Code

In this step, cloud developers make implementation decisions that best fit the individual requirements of each microservice included in an increment, and then complete the previously generated *Increment Cloud Artifacts Model* by specifying: (i) the technology in which *Artifacts* included in a *DeploymentProject* will be implemented; (ii) inter-service communication information of *Implementation/InteractionProjects* (e.g., SOAP/REST service style, message format, protocols) along with configuration information of *HostedServices* that will change at runtime, by creating or updating classes of type *DynamicConfiguration*, *Setting* or *EndPoint*; (iii) the representation of *Artifacts* (e.g., source code language); and (iv) the location where the *Artifacts* will be generated. Next developers execute M2T transformations that use this model and the *Extended Increment Architecture Model* as input in order to generate cloud *Artifact* implementations, which are organized into a directory structure according to the *Location* specified for each *Project*. The generated cloud *Artifacts* implement (see Fig. 1): (i) *Interaction Protocols* (e.g., choreography),

```
01.   rule ParticipantUse2Implementation {
02.   from
03.       ParticipantUseInput : EIAM!ParticipantUse (
04.           ParticipantUseInput.scalability = #HighVolumenRequests)
05.   to
06.       ImplementationProject : CAM!ImplementationProject (
07.           name <- ParticipantUseInput.name,                          -- assign the Participant name to the Project name
08.           artifactImpact <- ParticipantUseInput.architecturalImpact,  -- propagate architectural impact values
09.           deployment <- Deploy,
10.           artifacts <- Set{ImplementationProjectConfiguration,
11.               FrontEndConfiguration, FrontEndConfiguration,
12.               ParticipantUseInput.role ->select(e | e.isConsumer)->collect(e | thisModule.Role(e))},
13.           services <- FrontEndService,
```

Fig. 4 Excerpt of M2M for generating the *Increment Cloud Artifact Model*

(ii) software *Cloud Adaptors,* (iii) skeletons of microservices logic, *Interfaces,* client objects that invoke services and initiate interaction, APIs that microservices expose, and as many configuration files as *DynamicConfiguration Environments* (e.g., development, production), (iv) *Building/Packaging Scripts* to create deployable packages, according to the *DeploymentProjects'* structure. Finally, cloud developers complete the generated cloud *Artifacts* and execute the packaging/building scripts obtaining deployable packages.

3.3   Deployment and Architectural Reconfiguration

In this activity architects select the adaptation patterns best suited to integrating the increment's architecture into the current application architecture. Additionally, architects make provisioning and deployment decisions about the infrastructure and platform resources that must be provisioned in order to deploy the microservices included in a deployment artifact, then execute M2T transformations that generate cloud artifacts that operationalize the adaptation patterns according to *Extended Increment Architecture Model* and the *Increment Cloud Artifacts Model.* The cloud artifacts generated are (see Fig. 1): (i) *Deployment Scripts* with which to deploy (and provision) previously generated packages along with the corresponding configuration files, and (ii) scripts with which to reconfigure the application architecture, which use architectural impact specification to dynamically update *EndPoints* information stored in the microservice configuration files. For deeper information about how to document provisioning and deployment decisions, as well as about the generation of deployment scripts see [21].

Finally, the *Extended Increment Architecture Model* and the *Increment Cloud Artifacts Model* are used as the input for M2M transformations that update the current *Application Architecture Model* and the *Application Cloud Artifacts Model* by integrating the corresponding architectural elements and cloud artifact descriptions.

The *Increment Implementation* and *Deploy and Architectural Reconfiguration* activities allow developers to satisfy the *Decentralized governance* and *Infrastructure Automation* microservice principles by providing models that abstract implementation and deployment decisions from technological aspects, and tools that enable developers to obtain software artifacts that can be used as part of CI/CD pipelines.

4   Case Study

In order to illustrate the use of our approach, in this section we present an excerpt of a case study (adapted and extended from [3]). A manufacturing company wishes to improve the technological support given to its dealers, and is considering updating

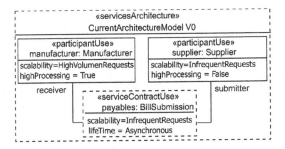

Fig. 5 Excerpt of the current *Application Architecture Model*

its already existing manufacturer microservice by including new functionalities with which to allow dealers to place production orders and obtain the products ordered by means of a shipping service. Figure 5 shows an excerpt of the current *Application Architecture Model* which will evolve after integrating the Manufacturer's microservice update.

The development team involved in this new requirement used SoaML to model the architectural design of the new manufacturer microservice functionalities and produced the *Microservice Architecture Model* (Fig. 6), described as a *Services Architecture*, whose inner parts (e.g., *ServiceContracts, Interfaces, Roles*) are not shown owing to space restrictions. *The Microservice Architecture Model* includes microservice architectural elements that describe microservice logic as well as microservice interoperation requirements (depicted with a background color in Fig. 6). Note that the *Participants* that are expected to interoperate with the manufacturer microservice (other components/microservices that consume manufacturer microservice's services or provide it with services—*outside Roles*) are indicated by *ParticipantUses* with dashed outlines (i.e., *:Dealer* and *:Shipper*), whereas that internal microservice components are indicated by *ParticipantUses* with continuous outlines (i.e., *:Fulfilment* and *:Production*).

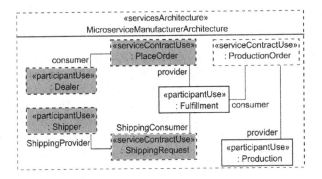

Fig. 6 Excerpt of the *Microservice Architecture Model*

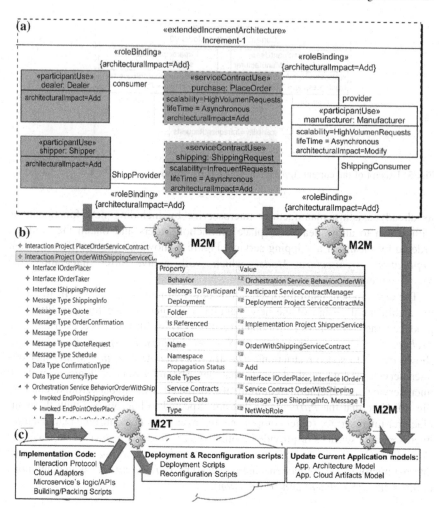

Fig. 7 Main transformation chains **a** *Extended Increment Architecture Model*, **b** *Increment Cloud Artifacts Model*, **c** generated cloud artifacts and updating of current application models

Figure 7a shows the *Extended Increment Architecture Model* resulting from the *Increment Integration Specification* activity, in which the *Microservice Architecture Model* (Fig. 6) becomes the *Service Architecture of the participant Manufacturer*, which is referenced by the *Manufacturer ParticipantUse*. The microservice inter-operation requirements described in the *Microservice Architecture Model* (depicted with a background color in Fig. 6), were referenced in the Extended *Increment Architecture Model*, thus becoming the microservice integration logic (depicted with a background color in Fig. 7a).

Architects proceed to specify the *Participants* that will play the roles defined in the integration logic. The *Participant Manufacturer* already exists in the *Current Architecture Model* and its implementation will change in order to implement the interfaces required to play the *provider Role*, thus it is tagged with *architectural Impact = Modify*. The *Dealer* and *Shipper* participants do not exist in the *Current Architecture Model* and must be created, therefore they are tagged with *architecturalImpact = Add*. Finally, architects analyze the nature of the work that *Participants* and *ServiceContracts* will perform. The *PlaceOrder* service is expected to be highly demanded, and the *ParticipantUse Manufacturer* plays the *Role* of *provider* then it is tagged with values *scalability = HighVolumenRequests*, and *lifetime = Asynchronous*. The *ServiceContractUse purchase:PlaceOrder* will manage the interoperation, then it is tagged with equal values (see Fig. 7a).

During the *Increment Implementation* activity there were no inconsistencies among *Participants'* interfaces, and the interaction protocols described in the integration logic were not therefore changed. The *Increment Artifacts Model* (Fig. 7b) was automatically obtained by defining and executing M2M transformations in the Eclipse extension Atlas Transformation Language (ATL), then it was completed. Skeletons of source code that implement microservices logic (see Fig. 7c) were obtained by defining and executing M2T transformations in the Eclipse extension Acceleo. Once skeletons were completed we built the application, packed it and deployed it in the Microsoft Azure cloud environment. Figure 8 shows an excerpt of the transformation rule applied to generate skeletons of source code that implement *Interfaces* corresponding to the *Roles* (lines 12, 13) played by a microservice. Visual Studio compatible files (line 13) were generated for each new microservice (line 8) that consume another microservice service (line 10).

During the *Deployment and Architectural Reconfiguration*, we use the open source Eclipse extension Acceleo M2T generator in order to obtain *Reconfiguration Scripts* (see Fig. 7c). We generated XML Document Transform (XDT) files used in Visual Studio to modify service configuration files while the deployment takes place. Finally, the M2M transformations that update current application models (see Fig. 7c) are in the process of being built; however Fig. 9 shows how the *Application Model Architecture* is expected to look after integration.

```
01.   [template public generateElement(anExtendedIncrementArchitectureModel : ExtendedIncrementArchitectureModel)]
02.   [comment @main/]
03.
04.   [for(EIA:ExtendedIncrementArchitecture | self.collaboration->
05.     select(oclIsTypeOf(ExtendedIncrementArchitecture)))]
06.     [for(SCU:ServiceContractUse | EIA.servicecontractuses)]
07.       [for(RB:RoleBinding | SCU.rolebinding2->
08.         select(e:RoleBinding | e.architecturalImpact = ArchitecturalImpact::Add))]
09.         [for(R:Role | RB.client->select(oclIsTypeOf(Role)))]
10.           [if (R.isConsumer)]
11.             [if (R.roleType->notEmpty())]
12.               [let inter:Interface = R.roleType]
13.                 [file (inter.name.concat('.svc.sc'), false)]
```

Fig. 8 Excerpt of M2T used to generate *Reconfiguration Scripts*

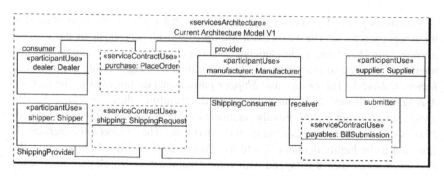

Fig. 9 Current *Application Architecture Model* after integration

5 Conclusions and Future Work

We presented a general view of a method for the incremental integration of microservices into cloud applications. In this method, developers specify how to integrate a microservice into the current application by describing both the integration logic and the architectural impact of integration without taking into consideration the specifics of any cloud environment. They then use both the microservice design and the integration specification to generate: (i) source code that implement skeletons of microservice's logic as well as integration logic, (ii) scripts to build and package the related microservice software artifacts, (iii) scripts to deploy the microservices, and (iv) scripts to manage the current application's architectural reconfiguration produced by the integration. Particular emphasis has been placed on explaining how the method manages to keep the microservice design independent from the integration specification, thus allowing different development teams to work on different microservices and giving them the independence to design, implement and deploy microservices according to the implementation/deployment technological requirements of each microservice. Providing developers with tools that automate integration and deployment operations help developers in eliminating discontinuities between development and deployment through CI/CD support which is required in order to deliver new functionalities to customers in an agile manner.

We have shown the feasibility of our proposal by applying it to a case study. We are currently working on implementing transformation chains; however, our approach does not take into account the automation of infrastructure changes. We are considering the use of the DevOps approach in order to improve the collaboration between development and operations, thus allowing new software releases to be made available much faster [25]. In this context, as further work we plan to adapt the method presented in this work in order to satisfy DevOps practices which promote the automation of the process of software delivery and infrastructure changes. Additionally, even though the offered models allow version control and we propose to generate software artifacts according to the architectural impact of integrating microservices, the approach to generate software artifacts that propagate design decisions related to updating or deleting already deployed microservices needs to be implemented. We also plan to provide mechanisms to manage incremental consistency, avoiding to lose changes introduced in the implementation code after generation (e.g., changes in interface implementations).

We identified some limitations, architectural reconfiguration is achieved by deploying/redeploying/undeploying microservices and by updating binds among them; however, we are not managing the updating of running instances of microservices. This is a challenging task, since cloud providers offer some proprietary instance management functionalities. Fortunately, the model-driven approach followed by our method enables us to abstract the instance management mechanisms, as well as to describe some proprietary advanced characteristics at a detailed level. Finally, we also plan to design experiments with which to validate the effectiveness of our approach in practice.

Acknowledgements This research is supported by the Value@Cloud project (MINECO TIN2013-46300-R), DIUC_XIV_2016_038 project, and the Microsoft Azure Research Awards.

References

1. Feitelson, D.G., Frachtenberg, E., Beck, K.L.: Development and deployment at facebook. IEEE Internet Comput. **4**, 8–17 (2013)
2. Familiar, B.: Microservices, IoT, and Azure: Leveraging DevOps and Microservice Architecture to Deliver SaaS Solutions. Apress (2015)
3. Fowler, M., Lewis, J.: Microservices: a definition of this new architectural term. http://martinfowler.com/articles/microservices.html
4. Newman, S.: Building Microservices. O'Reilly Media, Inc. (2015)
5. Hillah, L.M., Maesano, A., De Rosa, F., Maesano, L., Lettere, M., Fontanelli, R.: Service functional test automation. In: 10th Workshop on System Testing and Validation. Sophia Antipolis (2015)
6. Balalaie, A., Heydarnoori, A., Jamshidi, P.: Migrating to cloud-native architectures using microservices: an experience report, pp. 1–15 (2015)
7. Chow, R., Golle, P., Jakobsson, M., Shi, E., Staddon, J., Masuoka, R., Molina, J.: Controlling data in the cloud: outsourcing computation without outsourcing control. In: Proceedings of the 2009 ACM Workshop on Cloud Computing Security, pp. 85–90 (2009)
8. Krylovskiy, A., Jahn, M., Patti, E.: Designing a smart city internet of things platform with microservice architecture. In: 2015 3rd International Conference on Future Internet of Things and Cloud, pp. 25–30 (2015)
9. Stefan, B.: How we build microservices at karma. https://blog.yourkarma.com/building-microservices-at-karma
10. Frey, S., Hasselbring, W.: The cloudMIG approach: model-based migration of software systems to cloud-optimized applications. Int. J. Adv. Softw. **4**, 342–353 (2011)
11. Guillén, J., Miranda, J., Murillo, J.M., Canal, C.: Developing migratable multicloud applications based on MDE and adaptation techniques. In: Proceedings of the Second Nordic Symposium on Cloud Computing and Internet Technologies—Nordic '13, pp. 30–37 (2013)

12. Ardagna, D., Di Nitto, E., Casale, G., Petcu, D., Mohagheghi, P., Mosser, S., Matthews, P., Gericke, A., Ballagny, C., D'Andria, F., et al.: MODAC LOUDS : a model-driven approach for the design and execution of applications on multiple clouds. In: Proceedings of the 4th International Workshop on Modeling in Software Engineering, pp. 50–56 (2012)
13. Zuñiga-Prieto, M., Abrahao, S., Insfran, E.: An incremental and model driven approach for the dynamic reconfiguration of cloud application architectures. In: 24th International Conference on Information Systems Development ISD2015 (2015)
14. Zuñiga-Prieto, M., Gonzalez-Huerta, J., Abrahao, S., Insfran, E.: Towards a model-driven dynamic architecture reconfiguration process for cloud services integration. In: 8th International Workshop on Models and Evolution (ME 2014) co-located with ACM/IEEE 17th International Conference on Model Driven Engineering Languages and Systems, pp. 52–61. Valencia, Spain (2014)
15. Viktor, F.: The DevOps 2.0 Toolkit: Automating the Continuous Deployment Pipeline with Containerized Microservices. CreateSpace Independent Publishing Platform (2016)
16. Vijaya, A., Neelanarayanan, V.: Framework for platform agnostic enterprise application development supporting multiple clouds. Procedia Comput. Sci. **50**, 73–80 (2015)
17. Bergmayr, A., Troya, J., Neubauer, P., Wimmer, M., Kappel, G.: UML-based cloud application modeling with libraries, profiles, and templates. In: CloudMDE@ MoDELS, pp. 56–65 (2014)
18. Guillén, J., Miranda, J., Murillo, J.M., Canal, C.: A UML Profile for modeling multicloud applications. In: European Conference on Service-Oriented and Cloud Computing, pp. 180–187 (2013)
19. Brandtzæg, E., Mosser, S., Mohagheghi, P.: Towards CloudML, a model-based approach to provision resources in the clouds. In: 8th European Conference on Modelling Foundations and Applications (ECMFA), pp. 18–27 (2012)
20. Brogi, A., Ibrahim, A., Soldani, J., Carrasco, J., Cubo, J., Pimentel, E., D'Andria, F.: SeaClouds: a European project on seamless management of multi-cloud applications. ACM SIGSOFT Softw. Eng. Notes **39**, 1–4 (2014)
21. Zúñiga-Prieto, M., Insfran, E., Abrahão, S.: Architecture description language for incremental integration of cloud services architectures. In: IEEE 10th Symposium on the Maintenance and Evolution of Service-Oriented Systems and Cloud-Based Environments (MESOCA), Raleigh, USA (2016)
22. Object Management Group: Service oriented architecture Modeling Language (SoaML) Specification. http://www.omg.org/cgi-bin/doc?formal/2012-03-01.pdf (2012)
23. Costa, B., Pires, P.F., Delicato, F.C., Merson, P.: Evaluating REST architectures-approach, tooling and guidelines. J. Syst. Softw. **112**, 156–180 (2014)
24. Humble, J., Farley, D.: Continuous Delivery: Reliable Software Releases through Build, Test, and Deployment Automation. Pearson Education (2010)
25. Wettinger, J., Andrikopoulos, V., Leymann, F.: Enabling DevOps collaboration and continuous delivery using diverse application environments, pp. 348–358 (2015)

Browsing Digital Collections with Reconfigurable Faceted Thesauri

Joaquín Gayoso-Cabada, Daniel Rodríguez-Cerezo
and José-Luis Sierra

1 Introduction

Faceted navigation is a common interaction technique in business, the cultural industry and many other domains [2, 18, 26, 29, 30, 32]. For this purpose, resources are classified in terms of suitable *faceted thesauri*. A faceted thesaurus groups classification terms into facets, which in turn can have associated sub-facets, yielding a hierarchical arrangement. This hierarchical organization can, in turn, guide navigation through the underlying collection of digital resources (regardless of whether these are records in a database, objects in a virtual museum, entries in a virtual shop catalog, or any other type of digital object).In mature digital collections, in which there are few or no changes in the underlying resources, and in which classification schemata are pre-established and stay immutable, faceted navigation can be accomplished in very efficient ways [7]. However, for live collections, such as those arising in social or other highly dynamic and changing environments, not only are changes in the underlying resources frequent, but these changes can also affect the classification schemata themselves. When faceted thesauri are used in these dynamic settings, reconfiguring the thesaurus can mean a profound rearrangement of the collection's internal structures, which can be costly

A prior version of this paper has been published in the ISD2016 Proceedings (http://aisel.aisnet.org/isd2014/proceedings2016).

J. Gayoso-Cabada · D. Rodríguez-Cerezo · J.-L. Sierra (✉)
Complutense University of Madrid, Madrid, Spain
e-mail: jlsierra@ucm.es

J. Gayoso-Cabada
e-mail: jgayoso@ucm.es

D. Rodríguez-Cerezo
e-mail: drcerezo@ucm.es

© Springer International Publishing Switzerland 2017
J. Gołuchowski et al. (eds.), *Complexity in Information Systems Development*,
Lecture Notes in Information Systems and Organisation 22,
DOI 10.1007/978-3-319-52593-8_5

69

in time (often, it must be carried out offline). In consequence, user experience can be seriously hindered. Indeed, when a user changes the thesaurus, what he/she probably expects is an almost instant response in navigation; in these cases, high response times and/or a temporarily outdated underlying information system are inadmissible. We have realized this fact during the compilation of research and education-oriented collections of digital objects in digital humanities scenarios [4, 23, 24]. In these scenarios faceted thesauri-like classification schemata were subjected to continuous change, refinement and evolution throughout the collections' life cycles. Many times those reconfigurations in the schemata were performed with experimental and/or exploratory purposes in mind, and domain experts (researchers and/or instructors in charge of compiling and maintaining the collections) were not willing to wait for long periods until the changes were reflected in their collections. On the contrary, they wanted to see the changes in the browsing system immediately after changing the classification schemata, in order to determine whether these changes in the schemata really met their expectations. Thus, in this paper we partially respond to these needs by firstly providing a model of digital collection with a reconfigurable faceted thesaurus, in which the facet hierarchy can be freely rearranged, thus accomplishing the exploratory needs of the potential users. Secondly, we also introduce indexing strategies that provide reasonable time-space tradeoffs concerning navigation reconfigurability, while preserving acceptable levels of user experience.

The rest of the paper is organized as follows. Section 2 describes the digital collection model. Section 3 addresses browsing in the presence of the kind of reconfigurable thesauri introduced by this model. Section 4 introduces some works related to our browsing approach. Finally, Sect. 5 outlines the final conclusions and some lines of future work.

2 Digital Collections with Reconfigurable Faceted Thesauri

In this section we introduce our model of digital collection with reconfigurable thesauri. Section 2.1 describes the structure of these collections. Section 2.2 addresses thesauri reconfiguration.

2.1 Structure of the Collections

Our collections comprise the following parts (see Fig. 1 for an example):

- On one hand, there are the *resources* in the collection. These resources are digital objects whose nature is no longer constrained by the model. Thus, these resources can be media files (images, sound, video, etc.), external resources

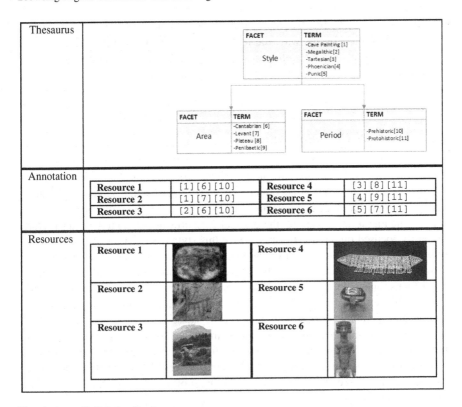

Fig. 1 A small digital collection

identified by their URIs, or entities of a more abstract nature (tuples of a table in a relational database, records in a bibliographical catalog, elements in an XML document, rows in a spreadsheet, etc.). For instance, the small collection depicted in Fig. 1 includes six image archives as resources, corresponding to photographs of artistic objects from the Prehistoric and Protohistoric artistic periods in Spain (Fig. 1 actually shows thumbnails of these images).

– On another hand, there is the *annotation* of the resources. This annotation consists of associating descriptive *terms* with resources. These terms are useful when cataloguing resources and, therefore, they enable future uses of the collection (navigation, search, etc.). Since each term has a unique identifier associated, annotating a resource consists of associating a set of term identifiers with such a resource. For instance, in Fig. 1 resource number 1 has the terms identified by **[1] [6] [10]** associated.

– Finally, there is a faceted *thesaurus* that organizes the terms into facets and which arranges these facets hierarchically. For instance, the faceted thesaurus in Fig. 1 includes a root facet *Style*, representing the artistic style used, and two sub-facets: *Area* (representing the geographical area), and *Period* (representing the artistic period). Each facet includes representative terms related to this facet.

Notice how each term consists of a descriptive name and the aforementioned unique identifier. In this way, the terms indicated below ([1] [6] [10]) actually refer to the terms *Cave Painting* in the facet *Style*, *Cantabrian* in the facet *Area*, and *Prehistoric* in the facet *Period*.

2.2 Thesauri Reconfiguration

Our model lets users reconfigure thesauri by rearranging the hierarchical organization of facets in order to accommodate their experimental and/or exploratory needs. For instance, Fig. 2 shows an example concerning the collection in Fig. 1. Indeed, the thesaurus in Fig. 2a (the original thesaurus in Fig. 1) reflects a structure primarily focused on the artistic style. Once this style has been set, it is possible to introduce either a geographical or an artistic period refinement. However, it may also be feasible to conceive of an alternative organization, with the artistic period as main focus, and with the geographical area and style as secondary features. This leads to the thesaurus in Fig. 2b, which has been obtained from the original one by altering the hierarchical facet arrangement.

Since the organization of a collection ultimately relies on its faceted thesaurus, by reconfiguring this thesaurus it is possible to implicitly reconfigure the structure of the entire collection, adapting it to different use scenarios as needed. This effect can be readily appreciated on the *navigation map* of a collection. Such a map is a directed graph in which:

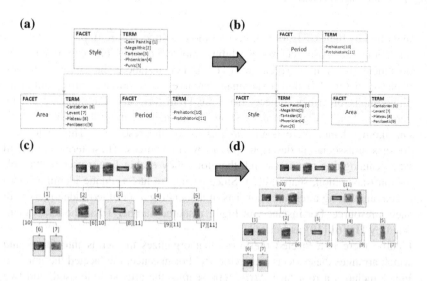

Fig. 2 a Original thesaurus in Fig. 1; **b** Reconfigured thesaurus; **c** Navigation map induced by (**a**); **d** Navigation map induced by (**b**)

- Nodes represent sets of resources, and arcs are labelled with terms used to narrow down the resources in the source nodes in order to yield the resources in the target ones (actually, all those resources in the source nodes annotated with the terms in the arcs).
- Structure is constrained by the facet hierarchy. In this way, root nodes can only be narrowed down with terms in root facets, and, if a node is produced by a term in a facet, it can only be narrowed down with terms in sub-facets of the mentioned facet.

Thus, reconfigurations in the thesaurus affect the entire navigation map. It is made apparent in Fig. 2c, d, which, respectively, outline navigation maps of the collection in Fig. 1 before (Fig. 2c) and after (Fig. 2d) thesaurus reconfiguration.

3 Browsing with Reconfigurable Faceted Thesauri

As the previous section makes apparent, reconfigurations in the hierarchical structure of a thesaurus profoundly impact the structure of the overall collection. In particular, after reconfiguration, the collection's navigation map can be completely altered. This hampers the use of efficient implementations of faceted browsing (e.g., [7]), which are basically driven by the navigation map structure and therefore require pre-established and unmodifiable faceted thesauri (otherwise, the navigation map would have to be regenerated after each thesaurus' reconfiguration, which would be a costly task even for collections of moderate size, and which could seriously impact user experience). Thus, by allowing reconfigurability, it is necessary to switch to alternative representations, enabling *all* the possible navigations induced by *all* the possible reconfigurations of the faceted thesaurus. Section 3.1 characterizes the expected behavior as a finite state machine. Section 3.2 addresses some complexity issues associated with a naïve representation directly based on such a machine. Finally, Sect. 3.3 discusses some indexing approaches that we have used to face these complexity issues.

3.1 Navigation Automata

In order to characterize all the possible navigations induced by all the possible reconfigurations of a faceted thesaurus it is necessary to free terms from the faceted structure. Therefore, a plain set of terms must be considered and, in each interaction state, all the meaningful selection of terms must be applied. The result can be represented as a finite state machine that we will call a *navigation automaton*. This automaton will consist of *states* labelled by sets of resources, and *transitions* labelled by terms. More precisely:

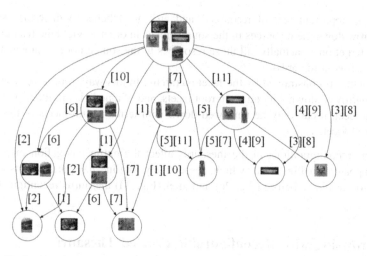

Fig. 3 Navigation automaton for the collection in Fig. 1

- There will be an initial state labelled by all the resources in the collection.
- Given a state S labelled by a set of resources R, for each term t annotating some resource in R there will be a state S' labelled by all the resources in R annotated by t, as well as a transition from S to S' labelled by t.[1]

Figure 3 shows the navigation automaton for collection in Fig. 1. Notice that the navigation automaton does not depend on the hierarchical organization of facets in the thesaurus, but only on the terms and on the resources in the collection. Therefore, it is not affected by reconfigurations in the thesaurus. Indeed, it can be thought of as the amalgam of all the possible navigation maps induced by all the possible reconfigurations of the collection thesaurus.

Since the navigation automaton embeds all the possible navigation maps, faceted browsing with respect to a particular thesaurus configuration can be formulated in a straightforward way, since there will be a direct correspondence among interaction states in the browsing process and states in the navigation automaton. In addition, in each interaction state will be a set of allowable facets to be explored. Indeed:

- The browsing process will start by considering the navigation automaton's initial state and the thesaurus' root faces as allowable ones.
- In each interaction state, the allowable facets will be used to constraint the possible terms to continue browsing. Once an allowed term is selected, the navigation automaton will be used to establish the new navigation state and the thesaurus to update the allowable facets.

---

[1]Notice that S and S' can be the same—when all the resources in R are annotated by t.

3.2   Complexity Issues

As indicated in the previous subsection, the explicit availability of the navigation automaton provides an elegant and efficient solution to faceted browsing in the presence of a reconfigurable thesaurus. Unfortunately, for collections with dense annotations there is the risk of facing unacceptably growing rates in the number of resulting states. It should not be surprising since we are attempting to represent all the possible ways of navigating in a single structure, regardless of the structure of the underlying thesaurus. In the worst case, the number of states can grow exponentially with respect to the number of resources. This extreme case, in which the number of states is 2^n-1 (with n the number of resources), arises, for instance, by distinguishing each pair of resource annotations in a single term (Fig. 4 shows an example with 4 resources and 4 terms).

While the extreme case presented can be somewhat artificial, it cannot be ignored if we hope to deal with arbitrary evolving collections. For this purpose, it can be desirable to look for alternative indexing approaches to enable the dynamic recreation of the relevant parts of the navigation automaton during browsing while preserving required levels of user experience.

3.3   Indexing Strategies

In order to deal with the complexity issues raised in the previous subsection, we have explored two different indexing strategies: *inverted indexes* and *navigation*

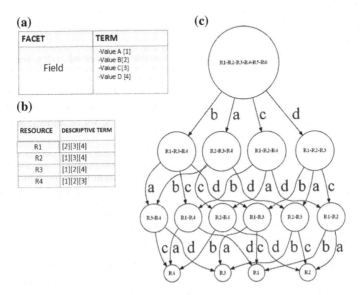

Fig. 4 Example of exponential explosion: **a** a simple thesaurus, **b** a simple associated collection, **c** the resulting navigation automaton

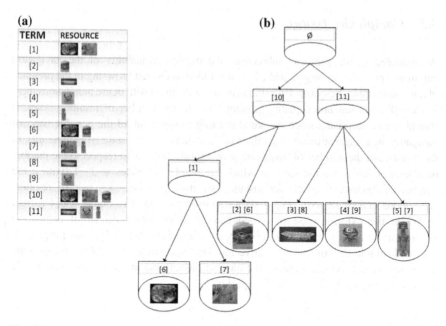

Fig. 5 **a** An inverted index for the collection in Fig. 1a; **b** a navigation dendrogram for such a collection

dendrograms. Following paragraphs analyze these strategies and provides some empirical comparison results.

Inverted Indexes

Inverted indexes are standard artifacts used for information retrieval [33]. Basically, for each description term, an inverted index associates the set of resources annotated with this term. Figure 5a shows an example of inverted index for the collection in Fig. 1.

Inverted indexes can be used to recreate the browsing behavior described in the previous section in a straightforward way. Basically, it suffices to maintain interaction states consisting of the set of terms chosen and the set of allowable facets to be explored.

- The initial state is constituted by the empty set of terms and by the root facets.
- Given an interaction state the selection of the next term to be applied obeys the same constraints as with the navigation automaton: this term must be a term t included in some facet f among the allowable facets. Then, the new interaction state will consist of: (i) all the terms in the previous one plus the new term selected, (ii) all the sub-facets of f.

Concerning the resources filtered in each interaction state, these resources can be determined by considering the set of terms $\{t_1, ..., t_n\}$ in such a state and by evaluating the conjunctive query $t_1 \wedge ... \wedge t_n$ using the inverted index.

The cost of evaluating the queries $t_1 \wedge \ldots \wedge t_n$ in each interaction state constitutes the main shortcoming of the approach. Indeed, it involves finding the intersections of the sets for $t_1 \ldots t_n$ in the inverted index. While there has been extensive research in performing these intersection operations efficiently [5], the cost is not negligible.[2] On the positive side is the availability of many mature implementations and frameworks that can be used in a straightforward way to support the technique. For instance, in our experiences, we used Lucene [17] for such a purpose.

Navigation Dendrograms

In order to avoid the proliferation of intersection operations, which is characteristic of inverted indexes representations, we have envisioned a tree-shaped indexing scheme inspired by *dendrograms* in hierarchical clustering [13]. The resulting representations are called *navigation dendrograms*. Following hierarchical clustering principles, nodes in the dendrogram represent subsets of the overall resource set. In this way:

- The dendrogram's root represents the whole resource set.
- If a node represents a particular resource set, then each child node represents a partition of this set (i.e., child nodes represent mutually disjoint subsets of the parent's set).

The resource set associated to a node is not explicitly stored in this node. Instead, each resource is only hosted in one node (the resource's *host node*). Resources placed in a node are called the mentioned node's *own resources*. The overall resource set of a node is given by its own resources and by all the own resources of its descendants. Finally, in order to partition the resource space, each node has a set of *filtering terms* associated, so that all the own resources in the node and in all their descendants' must be annotated with these filtering terms (the node is said to *filter* those resources). Figure 5b shows a navigation dendrogram for the collection in Fig. 1.

Initially, the dendrogram contains a single root node with an empty filtering set. Then, the dendrogram is incrementally constructed by sequentially adding resources, one resource at a time. Pseudocode in Fig. 6 details how a new resource is added to the dendrogram. Basically the process proceeds by looking for a host node for the resource, creating new nodes when needed. For this purpose the resource is firstly filtered through the dendrogram's nodes. When a node with no children filtering the resource is reached:

- If the resource has some term not included in any of the filtering term in the traversed nodes, and there is a child node containing some of these non-considered terms in its filtering set, a new intermediary *fork* node is created

---

[2]Notice that, although as indicated earlier, frameworks like Solr [7] support faceted browsing in a straightforward and efficient manner by identifying paths in the thesaurus with terms, in our context these features are useless, since thesauri can be reconfigured anytime, thus invalidating this solution. So we are confined to explicitly evaluating conjunctive queries in each interaction state.

```
Add resource r to dendrogram d:

CurrentNode = d's root
Δ_Res = r's terms

while there is some child n of CurrentNode such as
      n's filtering terms ⊆ Δ_Res {
   CurrentNode = One of such child nodes
   Δ_Res = Δ_Res - CurrentNode's filtering terms
}

InsertionNode = CurrentNode
if (Δ_Res ≠ ∅) {
   if there is some child n of InsertionNode such as
      n's filtering terms ∩ Δ_Res ≠ ∅ {
          ChildNode = One of such child nodes
          ForkNode = create new node
          Δ_Fork = Δ_Res ∩ ChildNode's filtering terms
          Δ_Child = ChildNode's filtering terms - Δ_Fork
          Δ_Res = Δ_Res - Δ_Fork
          set ForkNode's filtering terms to Δ_Fork
          set ChildNodes'filtering terms to Δ_Child
          change the arc InsertionNode → ChildNode
                  to InsertionNode → ForkNode
          add an arc ForkNode → ChildNode
          InsertionNode = ForkNode
   }
   if (Δ_Res ≠ ∅)) {
          HostNode =  create new node
          set HostNode's filtering terms to Δ_Res
          set HostNode's own resources to ∅
          add an arc InsertionNode → HostNode
          InsertionNode = HostNode
   }
}
add r to InsertionNode's own resources
```

Fig. 6 Pseudocode of the process for adding a resource to a navigation dendrogram

with this child node as child. In addition, the resource node is placed in the fork node or in a new host node (which is also included as a child of the fork node) depending on whether it has some term not included in the reached node's filtering set. In other case, the resource is also placed in a new host node, which is added as a new child node of the reached one.

– Otherwise, the resource is hosted in the reached node.

Thus, in the worst case, insertion of a resource involves the creation of two new nodes. In consequence, the number of nodes in the dendrogram is bound by $2R$ (with R the number of resources).

Concerning browsing, it is possible to conceive interaction states formed by:

– A set of *dendrogram nodes*, which are active in the interaction state.
– A set of allowable facets.

```
Find next interaction state of s given t of facet f:

notation: Given a dendrogram's node n:
    •  n↑: all the ancestors of n (including n itself)
    •  n↓: all the descendants of n (exluding n)
let N the set of dendrogram nodes in s {
    N' = ∅
    foreach n in N {
        if there is n' in n↑ such as t is in the filtering terms of n' {
            N' = N' ∪ {n}
        }
        else let  n↓t the nodes in n↓ with t in their filerting terms {
            N' = N' ∪ n↓t
        }
    }

    let F the set of subfacets of f {
    The next interaction state has N' as the set of dendrogram nodes
    and F as the set of allowable subfacets
    }
}
```

Fig. 7 Pseudocode to find the next interaction state during browsing

As in the earlier proposals, navigation firstly proceeds by selecting a term from one of the allowable facets. Then, the next interaction state can be obtained by:

– Establishing the allowable facets to the subfacets of the facet used.
– Refining the set of nodes according to the term selected. Nodes having such a term in their filtering terms or in the filtering terms of some ancestor are directly preserved. Nodes having this term in the filtering terms of some descendant are replaced by said descendants. Other nodes are discarded (see Fig. 7). Notice that, in order to speed up this computation, it is convenient to have direct access to the filtering terms for each node and its ancestors, as well as to have the node's descendants classified by their filtering terms (for the sake of simplicity, details are not shown in the pseudocode in Fig. 7).

Once the interaction state is determined, resources selected in an interaction state can be lazily recovered by iterating the dendrogram nodes and their descendants' own resource sets.

Finally, it is worthwhile to notice how navigation dendrograms overcome the main shortcoming of inverted indexes: the need to explicitly carry out set intersections during browsing. On the negative side, the indexing process is substantially more complex than in the case of inverted index construction.

Experimental Results

In order to compare the two indexing strategies described, we implemented both on *Clavy*, an experimental system for managing digital collections with reconfigurable

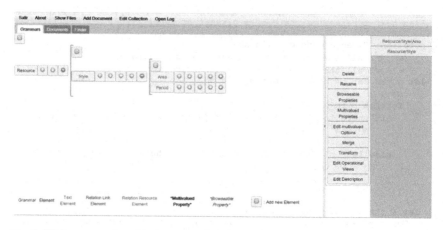

Fig. 8 Editing a reconfigurable thesaurus with Clavy

faceted thesauri-like schemata (see Fig. 8).[3] We also set up an experiment consisting of adding the resources in *Chasqui* [23],[4] a digital collection of 6283 digital resources on Precolombian American archeology, to *Clavy* and to simulate runs concerning browsing and schemata reconfiguration operations.

Each run was customized as follows. We interleaved resource insertion with browsing/reconfiguration rounds. Each insertion round consisted of 100 resource insertions (with the exception of the last one, in which the remaining resources where inserted). In turn, each browsing/reconfiguration round consisted of executing $0.1n$ browsing operations randomly interleaved with $0.01n$ reconfigurations (n being the number of resources inserted so far). Each browsing operation consisted, in turn, of selecting a feasible term, computing the next interaction state, and visiting all the resources filtered. Reconfiguration operations, then, consisted of feasible interchanges of two randomly selected facets,[5] followed by a browsing step.

Inverted indexes were managed using Lucene, while navigation dendrograms were managed using our own implementation (implemented in Java, as well as the Lucene framework). In both cases, in-memory indexes were used in order to avoid side effects of persistence, disturbing the experiment.

Figure 9 shows the results obtained from the two runs. The experiment was run on a PC with Windows 10, with a 3.4 GHz Intel microprocessor, and with 8 Gb of DDR3 RAM. The horizontal axis corresponds to the number of operations carried out so far. The vertical axis corresponds to cumulative time (in seconds). As is made apparent, the dendrogram-based approach clearly outperforms the inverted indexes (even though we are using a highly optimized framework, like Lucene, for inverted indexing vs. our own in-house experimental implementation for dendrograms).

---

[3]clavy.fdi.ucm.es/Clavy/.

[4]oda-fec.org/ucm-chasqui.

[5]By *feasible* we mean avoiding cycles in the resulting thesaurus.

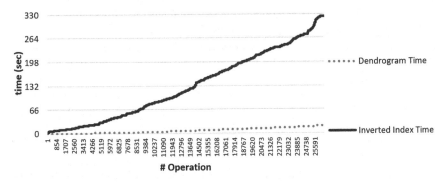

Fig. 9 Cumulative time of inverted indexes versus dendrograms

4 Related Work

There are several faceted browsing systems that, like ours, envision the possibility that the user reconfigures the underlying facet hierarchy. For instance, *mSpace* [22] makes it possible to organize information spaces in plain sets of facets, which can be interactively arranged in lists (*slices*), representing linear hierarchies between selected facets. Initially a slice may contain only a subset of the whole set of facets, and users can reorganize, add and delete the slice's facets according to their specific needs. In [21] an implementation based on semantic web (RDF) technologies is proposed. In [25] the advantages and disadvantages of this implementation are analyzed and more conventional solutions based on relational databases are proposed. In turn, */facet* [11] adopts a different approach as it is configured as a multi-faceted browser driven by RDFS schemata and RDF data. Facet hierarchies are inferred from RDF class hierarchies. In addition, being a multi-faceted browser, */facet* enables the user to jump from one facet hierarchy to another while maintaining filtering constraints. In consequence, actual facet hierarchies are not pre-established, but are dynamically configured according to the specific needs of each user. In addition, in [11] some efficiency issues associated with the extensive use of semantic web infrastructures are also reported. Contrary to *mSpace*, in which facets lack organization and in which users confer (linear) structures on these facets by arranging them in slices, or to */facet*, in which RDFS schemata pre-establish faceted hierarchies and in which the user is allowed to jump between hierarchies, in our approach, facets are doted with a hierarchical structure (the faceted thesaurus) from the beginning and users organize this hierarchy according to their needs. It provides users with more guidance during the reconfiguration process than *mSpace*'s approach and more freedom than */facet*'s (where reconfigurability consists of dynamically pasting different pre-established hierarchies). In addition, we propose efficient indexing approaches, specifically tailored to our model instead of piggybacking the implementation on general-purpose semantic web or relational database solutions.

Our navigation automaton model is actually equivalent to lattice-based proposals to browse information spaces, as described in the seminal work of [6]. In these proposals, resources are tagged with keywords. The lattice organization induced consists of nodes characterized by sets of resources and sets of keywords related by a *Galois connection* (i.e., the set of keywords is the intersection of the resources' keywords and the set of resources consists of all the resources filtered by the keyword set). This organization is actually the main subject of the fertile theory of *formal concept analysis* [20], where resources are called *objects*, keywords are called *attributes*, objects tagged with attributes are called *formal contexts*, and lattice nodes are called *formal concepts*. From this description, it should be apparent how states of navigation automata can actually be identified with formal concepts, and automata themselves with explicit representations of concept lattices (with an explicit representation of the whole order relation and an explicit labelling of the arcs with transition information). In [15] the intrinsic complexity of formal concept analysis is examined and the problem of determining the size of concept lattices is proved to be a #P-complete one (i.e., harder than NP-complete). In consequence, complexity results in concept lattice theory are directly translatable to navigation automata (in fact, construction of Sect. 3.2 was suggested by the proof of theorem 1 in [15]), In addition, there are several proposals on using concept lattices as the underlying indexing structures of digital collections. For instance, *ConceptCloud* uses concept lattices to support multilevel browsing of software repositories [10, 28] or academic publications [8]. A similar approach is described in [9], where concept lattices are used to organize the browsing of art-works from the University of Wollongong's Art Collection. All these proposals are also affected by the worst-case complexity of formal concept analysis.

Inverted indexes have been extensively used to support faceted browsing. In [31] the basic technique, as well as subsequent enhancements, are illustrated with the use of Lucene. In [27] an alternative approach, based on relational databases, is presented. However, all these approaches are based on the assumption of pre-established and immutable faceted thesauri. As noticed in [1], if this assumption is left out, and therefore arbitrary multilevel exploratory search is allowed, inverted indexes can become costly due to the set operations involved. For small amounts of terms, multidimensional structures (as used in data-warehouse and data-mining scenarios) can be advantageous [14]. However, the performance of these multidimensional approaches can dramatically decrease when dimensionality increases. For this purpose, in [1] a technique called *tree striping* is described, which proposes subdividing the overall information space in k disjoint sub-spaces, to apply standard inverted indexes or multidimensional indexing techniques to each resulting subspace and to use an efficient merging approach to aggregate the results of each subspace. Nevertheless, and contrarily to our navigation dendrograms, both multi-dimensional and tree striping techniques basically work with pre-established partitions of the information space, while in our approach thesauri are dynamic and evolving in nature.

Finally, it is worthwhile to notice that clustering techniques has been extensively used in social tagging systems (e.g., [12, 16, 19]) to enable the discovering of useful semantic relationships among tags in order to provide better guidance to users (e.g., by automatically discovering hierarchical structures of tags). Thus, clustering in these approaches is oriented to enhance users' browsing efficiency, while our navigation dendrograms are oriented to enhance the internal efficiency of the supporting software.

5 Conclusions and Future Work

Live digital collections, which involve active communities of specialized users (e.g., researchers or educators in a particular field), also require live organization schemata, which can be incrementally defined, refined and enhanced as collections evolve. In addition, in these scenarios users usually want systems to quickly respond to changes in the schemata, without waiting for costly and/or batch reorganization processes. In this paper we have addressed this problem of dynamic reconfigurability in the case of reconfigurable faceted thesauri, in which users can re-order facets in order to explore different and alternative ways of organizing the collections. Since facet hierarchy can be rearranged in unexpected ways, it is necessary to resort to a more free and exploratory browsing system. It has led us to model this system as a finite state machine, the *navigation automaton*, taking into account all the possible ways of navigating the collection, using terms selected from the facets. Unfortunately, we have also showed how, in some cases, the number of states in this automaton can increase exponentially with respect to the collection's size. In order to deal with this potential exponential factor we have explored two different indexing approaches: one based on standard inverted indexes (implemented in a robust and well-proven search framework: Lucene), and one inspired by hierarchical clustering techniques (the so-called *navigation dendrograms*). Some experiments with a real collection gave evidence of how the hierarchical clustering technique can outperform the inverted indexing one.

We are currently working on further optimizing our navigation dendrogram representation to leverage space requirements. Indeed, in order to provide efficient navigation we need to associate each node with the intersection and the union of all the terms annotating the resources under this node. In addition, for each union term we also need to store the descendant nodes that include such a term in their filtering sets. Fortunately, these sets present much regularity among nodes, which allows us to compress them by using tries and common node-set stores. Since the resulting structures provide time and space efficient representations for the nodes' intersection and union sets, all these optimizations enhance system performance, in addition to saving space. We are also looking for efficient ways to make all this information persistent, either by using standard relational databases or alternative NoSQL approaches (e.g., [3]), while causing again a minimum impact on system performance. Once efficient persistence mechanisms are established we want to run more empirical evaluations also taking persistence into account. We also hope to enhance our model with support for arbitrary Boolean queries and for different ways of exploring the resources selected. These mechanisms will be based on the navigation automaton model (supported by our indexing proposals, and, in particular, by navigation dendrograms) in order to get as much efficiency as possible. Finally, we plan to perform more comprehensive tests of our model in the context of different Digital Humanities efforts carried out by some of the Humanities research groups with whom we cooperate. Among these efforts we can highlight, in addition to the aforementioned *Chasqui* collection, different digital collections maintained by LEETHI, the UCM research group on European and Spanish Literatures, from Texts to Hypermedia (Mnemosine, a digital collection concerning rare texts from the Spanish Silver literature age,[6] Ciberia, a digital collection concerning Spanish digital literature[7] and Tropos, a digital collection concerning creative digital writing for literature education[8]), as well as those concerning the Panamanian "El Caño" archeological site,[9] created and curated by "El Caño" Foundation at Panama.

Acknowledgements This work has been supported by the BBVA Foundation (research grant HUM14_251) and by the Spanish Ministry of Economy and Competitiveness (research grant TIN2014-52010-R). The *Chasqui* repository was created and is maintained by Prof. Mercedes Guinea (currently a researcher at "El Caño" Foundation). The thesaurus used as an example in this work is adapted from the *Chasqui*'s cataloguing schema. *Chasqui*'s original software infrastructure was developed by Alfredo Fernández-Valmayor (currently also at "El Caño" Foundation).

References

1. Berchtold, S., Böhm, C., Keim, D.-A., Kriegel, H.-P., Xiaowei, X.: Optimal multidimensional query processing using tree striping. In: Proceedings of the 2nd International Conference on Data Warehousing and Knowledge Discovery, pp. 244–257. Springer, London, UK (2000)
2. Chengkai, L., Ning, Y., Senjuti, B-R., Lekhendro, L., Gautam, D.: Facetedpedia: dynamic generation of query-dependent faceted interfaces for wikipedia. In: Proceedings of the 19th International World Wide Web Conference, pp. 651–660. ACM, Raleigh, NC, USA (2010)
3. Chodorow, K.: MongoDB: the definitive guide. O'Reilly (2013)
4. Cigarrán-Recuero, J., Gayoso-Cabada, J., Rodríguez-Artacho, M., Romero-López, D., Sarasa-Cabezuelo, A., Sierra, J.-L.: Assessing semantic annotation activities with formal concept analysis. Expert Syst. Appl. **44**(11), 5495–5508 (2014)
5. Culpepper, J.-S., Moffat, A.: Efficient set intersection for inverted indexing. ACM Trans. Inf. Syst. 29(1), article 1 (2010)
6. Godin, R., Saunders, G.: Lattice model of browsable data space. Inf. Sci. **40**(2), 89–116 (1986)
7. Grainger, T., Potter, T.: Solr in Action. Manning Publications (2014)
8. Greene, G.-J., Dunaiski, M., Fischer, B.: Browsing publication data using tag clouds over concept lattices constructed by key-phrase extraction. In: Proceedings of Russian and South

---

[6]repositorios.fdi.ucm.es/mnemosine/.

[7]repositorios.fdi.ucm.es/CIBERIA.

[8]repositorios.fdi.ucm.es/Tropos/.

[9]oda-fec.org/nata.

African Workshop on Knowledge Discovery Techniques Based on Formal Concept Analysis, pp. 10–22. CEUR, Stellenbosch, South Africa (2015)

9. Greene, G.-J., Fischer, B.: Interactive tag cloud visualization of software version control repositories. In: Proceedings of the 3rd IEEE Working Conference on Software Visualization, pp. 56–65. IEEE, Raleight, NC, USA (2015)

10. Greene, G.-J.: A Generic framework for concept-based exploration of semi-structured software engineering data. In: Proceedings of the 30th IEEE/ACM International Conference on Automated Software Engineering, pp. 894–897. ACM, Lincoln, Nebraska, USA (2015)

11. Hildebrand, M., van Ossenbruggen, J., Hardman, L.: /facet: a browser for heterogeneous semantic web repositories. In: Proceedings of the 5th International Semantic Web Conference, pp. 272–285. Springer, Athens, GA, USA (2006)

12. Huang, J.-W., Chen, K.-Y., Chen, Y.-C., Yang, K.-N., Hwang, S., Huang, W.-C.: A novel spatial tag cloud using multi-level clustering. J. Inf. Sci. Eng. 30, 687–700 (2014)

13. Jain, A.-K., Murty, M.-N., Flynn, P.-J.: Data clustering: a review. ACM Comput. Surv. 31(3), 264–323 (1999)

14. Kriegel H.-P.: Performance comparison of index structures for multi-key retrieval. In: Proceedings of the ACM SIGMOD International Conference on Management of Data, pp. 186–196. ACM, Boston, MA (1984)

15. Kuznetsov, S.: On computing the size of a lattice and related decision problems. Order 18(4), 313–321 (2001)

16. Li, R., Shenghua, B., Fei, B., Su, Z., Yu, Y.: Towards effective browsing of large scale social annotations. In: Proceedings of 16th International World Wide Web Conference, pp. 943–952. ACM, Banff, Alberta, Canada (2007)

17. McCandless, M., Hatcher, E., Gospodnetic, O.: Lucene in Action, 2nd edn. Manning Publications (2010)

18. Perugini, S.: Supporting multiple paths to objects in information hierachies: faceted classification, faceted search, and symbolic links. Inf. Proc. Manag. 46(1), 22–43 (2010)

19. Radelaar, J., Boor, A.-J., Vandic, D., van Dam, J.-W., Fasinca, F.: Improving search and exploration in tag spaces using automated tag clustering. J. Web Eng. 13(3–4), 277–301 (2014)

20. Sarmah, A.-K., Hazarika, S.-M., Sinha, S.-K.: Formal concept analysis: current trends and directions. Artif. Intell. Rev. 44(1), 47–86 (2015)

21. Schraefel, M.-C., Smith, D-A., Owens, A., Russell, A., Harris, C., Wilson, M.: The evolving mSpace platform: leveraging the semantic web on the trail of the memex. In: Proceedings of the 16th Conference on Hypertext, pp. 174–183. ACM, Salzburg, Austria (2005)

22. Schraefel, M.-C., Wilson, M., Russell, A., Smith, D.-A.: mSpace: improving information access to multimedia domains with multimodal exploratory search. Commun. ACM 49(4), 47–49 (2006)

23. Sierra, J.-L., Fernández-Valmayor, A., Guinea, M., Hernanz, H.: From research resources to learning objects: process model and virtualization experiences. Education. Tech. Soc. 9(3), 56–68 (2006)

24. Sierra, J.-L., Fernández-Valmayor, A.: Tagging learning objects with evolving metadata schemas. In: Proceedings of the 8th IEEE International Conference on Advanced Learning Technologies, pp. 829–833. IEEE. Santander, Spain (2008)

25. Smith, D.-A., Owens, A., Schraefel, M-C., Sinclair, P., Max, P-A., Wilson, A., Rusell, A., Martinez, K., Lewis, P.: Challenges in supporting faceted semantic browsing of multimedia collections. In: Proceedings of the 2nd International Conference on Semantics and Digitial Media Technologies, pp. 280–283. Springer, Genoa, Italy (2007)

26. Tunkelang, D.: Faceted Search. Morgan & Claypool Publishers (2009)

27. Uddin, M.-N., Janecek, P.: The implementation of faceted classification in web site searching and browsing. Online Inf. Rev. 31(2), 218–233 (2007)

28. Way, T., Eklund, P.: Social Tagging for digital libraries using formal concept analysis. In: Proceedings of the 17th International Conference on Concept Lattices and their Applications, pp. 139–150. Sevilla, Spain (2010)

29. Wei, B., Liu, J., Zheng, Q.: A survey of faceted search. J. Web Eng. **12**(1–2), 41–64 (2013)
30. Yee, K.-P., Swearingen, K., Li, K., Hearst, M.: Faceted metadata for image search and browsing. In: Proceedings of the SIGCHI Conference on Human Factors in Computing Systems, pp. 401–408. ACM, Fort Lauderdale, Florida, USA (2003)
31. Yitzhak, O-B., Golbandj, N., Har'El N. et al.: Beyond basic faceted search. In: Proceedings of the 2008 International Conference on Web Search and Data Minining, pp. 33–44. ACM, Stanford, CA, USA (2008)
32. Zhang, Z., Li, W., Gurrin, C., Smeaton. A.-F.: Faceted navigation for browsing large video collection. In: Proceedings of the 22nd International Conference on Multimedia Modelling, pp. 412–417. Springer, Miami, USA (2016)
33. Zobel, J., Moffat, A.: Inverted files for text search engines. ACM Comput. Surv. **33**(2) (2006) article 6

E-Commerce Web Accessibility for People with Disabilities

Osama Sohaib and Kyeong Kang

1 Introduction

An e-commerce website is the central way an e-retailer communicates with their online consumers. E-retailer seeks to provide positive online purchasing experiences for online consumers of all ages. A website design encourages or discourages a consumer's online purchasing intentions [9]. In the context of business-to-consumer (B2C) e-commerce, website design features have different effects on forming consumers' trust and distrust [36]. In particular, B2C websites those are accessible content, information and easy to navigate influences consumer trust to buy online and must appeal to consumers [8, 11]. However, online shopping websites need to be accessible to all consumers of all ages, including those with disabilities. For-example, nowadays dynamic websites content (CSS—Cascading Style Sheets, Flash and JavaScript, etc.) are used in most of the e-commerce websites to provide a good visual presentation to attract or retain consumer. Though, these dynamic webpages are inaccessible to disabled people, such as visual impaired person, since screen readers are not adept of handling it. Many researchers such as [52, 53] have revealed dynamic contents using Flash animation, JavaScript, and graphic links etc. is a threat to web accessibility. The inability to shop online because of such interfaces increases web inaccessibility of e-commerce websites for people with disabilities.

A prior version of this paper has been published in the ISD2016 Proceedings (http://aisel.aisnet.org/isd2014/proceedings2016).

O. Sohaib (✉) · K. Kang
University of Technology Sydney, Sydney, Australia
e-mail: Osama.Sohaib@uts.edu.au

K. Kang
e-mail: Kyeong.Kang@uts.edu.au

© Springer International Publishing Switzerland 2017
J. Gołuchowski et al. (eds.), *Complexity in Information Systems Development*,
Lecture Notes in Information Systems and Organisation 22,
DOI 10.1007/978-3-319-52593-8_6

People with disabilities have limitations for going shopping, which put them at inconvenience because of their physical handicap. However, Internet has opened new possibilities of online shopping. In particular, disabled people can gain a sense of emotional stability by online shopping. Though, if e-commerce websites are inaccessible, consumers with disabilities do not have the equal access they are guaranteed by law. Many online consumers may have various types of disabilities, such as sensory (hearing and vision), motor (limited use of hands) and cognitive (language and learning disabilities) impairments. An accessible web site can utilize all of assistive technologies such as screen readers, voice recognition, alternative pointing devices, alternate keyboards, and the website displays [23]. Even though these technologies can help people with disabilities, webpage navigation, conveying image and flash based content are some of the serious issues for people with disabilities. Web accessibility is an important element in the design of e-commerce websites [24, 43]. Therefore, there is a serious need of web accessibility in B2C e-commerce websites for user of all ages and in particular for people with disabilities such as sensory (hearing and vision), motor (limited use of hands) and cognitive (language and learning disabilities) impairments.

The combination of technology and people in human computer interaction is a vital part of modern society that enables a wide range of economic benefits. In particular, the emerging growth of business-to-consumer (B2C) e-commerce allows everyone to put up his or her own business online, locally or globally. For-example, online shopping in Australia has become the new norm where more than 50% shop online [31]. According to 2012 digital media research, 75% of Australian made purchases from overseas online shops [39]. The web accessibility is an important factor that needs to be considered in Australian e-commerce [44]. Lazar and Sears [24], the authors discussed that web accessibility should receive attention in e-commerce websites. The B2C website provides the consumers with instant online access to products without physical barriers. In order to get the most out of revenue from online trade, businesses must focus on an accessible B2C e-commerce website, which should also give a real and convenient shopping experience for consumers of all ages. In particular, people with disabilities, such as color blindness. Because of the Internet availability, online shops provide consumer the ease of buying and selling products. However, the required web technological infrastructure is either insufficient or does not exist in terms of web accessibility guidelines. For that reason, the Web Content Accessibility Guidelines (WCAG 2.0) developed by the World Wide Web Consortium (W3C) help to make the website accessible for users of all ages and with disabilities such as colour blindness, deaf users, and age related vision problems. Therefore, the purpose of this study is to evaluate Australian B2C e-commerce website accessibility for consumers with disabilities in particular.

According to WHO [55] 2011 report 15% of the total population experience some form of disability. Although disability is a universal issue, it is particularly predominant in Australia. According to Australian Bureau of Statistics (ABS) Survey of Disability, Ageing and Carers, 2015 [56], 18.3% of Australians have a disability (Sight 30%, Hearing 10%). The research will be used as a guide for future improvement in e-commerce website accessibility. By assessing the web accessibility of online stores now, e-retailers may avoid legal problems in the future and possibly design accessible website. For that reason, this study analyses the Web Content Accessibility Guidelines (WCAG 2.0) in Australian B2C e-commerce websites using an online accessibility self-evaluation web service called "A-Checker" [1], to determine to what extent they meet the requirements of the web content accessibility guidelines (WCAG 2.0).

In particular, this study intends to focus on the following research questions. (1) What accessibility issues do Australian B2C e-commerce websites currently face? (2) What recommendations can be proposed in order to improve web accessibility in Australian B2C e-commerce websites? This study is organized as. The next section provides background and literature review. Then, research method is presented, followed by results and discussion. Finally, the study provides recommendations and concludes.

2 Background and Related Studies

2.1 Issues for Consumer with Disabilities in E-Commerce

As identified in the literature such as [51], the main disabilities that can hinder website accessibility are presented in Table 1.

The following examples may help to explain the types of barriers that can be encountered by an online consumer with a sensory (hearing and vision), motor (limited use of hands) and cognition (language and learning) disabilities.

Visual objects: Product images are placed on the e-commerce website. Websites that use flashing images could trigger symptoms for those with seizure disorders [15]. Flash animation makes web content inaccessible [52, 53]. These flash-based content or graphic images are an example of a barrier for consumers who are visually impaired. They are unable to see the product image and therefore unable to buy it. Such as the use of screen readers cannot read images, animations,

Table 1 Disabilities types and website accessibility issues [51]

Disability form	Symptoms'
Visual impairment	Partial vision; color blindness; may require usage of screen readers, or screen magnifier tools
Hearing impairment	Hearing difficulties; may require sound caption
Cognitive disability	Reading or comprehension difficulties; dyslexia; memory loss
Motor skills impairment	Inability to use keyboard/mouse; inability to make fine movements; may require usage of special assistive devices such as a voice browser, special joysticks and trackballs, and special keyboards that can be manipulated by fingers or using a head-wand

navigational buttons, as well as some difficulties with reading layout tables and charts [51]. In addition, color is perhaps one of the most used design elements to pass on information in websites. Therefore, a color-blind consumer (such as red-green color deficiency) making online purchase will find it difficult to process information and make purchasing decision based on color as a visual clue.

Audio objects: An e-retailer has recorded product information available to consumer on the website as an audio clip or digitized audio. Then the consumers who have difficulty in hearing or deaf cannot hear to buy the product.

Language difficulty: If the product information displayed on the website page is written in unnecessarily complicated language, such as the use of technical terms in specifications of electronics products. Then it may present serious difficulty for consumers with language, learning or cognitive disabilities (for-example, reading disabilities, thinking, remembering, sequencing disabilities). In addition, all consumers can face language comprehension barriers.

Objects Interaction: When an e-retailer focuses on more advanced image interactivity technology, such as three-dimensional (3D) virtual models, 360° rotation view, which allows the manipulation of product images. These will represent a barrier for a consumer who cannot grip the mouse to imitate actual experiences with the product. For-example, consumers with various forms of motor impairments may have increased difficulty using a mouse or keyboard. Such as, rollovers and drop-down menus are difficult to use without a mouse.

2.2 Web Accessibility

The World Wide Web Consortium (W3C) [49] defines web accessibility as "people with disabilities can perceive, understand, navigate, and interact with the web". People with disabilities include visual, speech, physical and neurological disabilities. There are varieties of web accessibility guidelines, but the most relevant are ISO 9241-151, Section 508 of the US rehabilitation Act and Web Content Accessibility Guidelines (WCAG) developed by the W3C. ISO 9241-151 and Sect. 508 comply with W3C. Web Content Accessibility Guidelines (WCAG) version 1.0 was developed in 1999, and then in 2008 more recommendations were made in WCAG 2.0 [50]. WCAG 2.0 is not limited to HTML but support various technologies [26].

The guidelines covered by WCAG 2.0 are: Perceivable, Operable, Understandable and Robust. The aim of "Perceivable" is to direct the user to perceive the user interface components. The "Operable" guides the users that how the interface should be operated and how to navigate. The "Understandable" means the web contents should be understandable by all users. The "Robust" describes that the information should be interpreted by the variety of users in the same way. WCAG 2.0 provided a testable success criterion for each accessibility guideline to determine whether a web page has met or failed the level of conformance. Three levels of conformance for web content accessibility guidelines (WCAG) are: Level A, Level AA and Level AAA.

Web developers must satisfy the Priority 1 (Level A) minimum level of conformance, Priority 2 is the Level AA includes all Levels of A and AA success criteria that the developer should satisfy. Priority 3 (Level AAA) is the highest level, the developer may satisfy the requirements for the webpage to help make accessible. These guidelines make web access easy for old age people and to people with disabilities. People who have cognitive and/or sensory, physical disabilities benefit from using accessible websites. The most common disabilities affecting the use of the website are hearing and visual impairment. In addition, people with difficulty gripping object, such as the use of a mouse requires additional access to web.

Methods for appraising web accessibility include automatic and expert evaluation. A number of researchers used automatic evaluation tools in their studies, such as in e-government, internet banking, schools, universities and company's website etc., for example [19, 20, 22, 25, 27, 35, 46, 47]. Other studies used a group of web accessibility experts for measurements, such as [3, 28, 37]. In addition, a number of researchers assessed web accessibility using both automatic and expert evaluation methods, such as [17, 18, 26, 34, 38, 41]. There are a various free web accessibility online tools examining web pages and automatically evaluating their compliance with Web Contents Accessibility Guidelines (WCAG), such as LIFT, Truwex, A-Prompt, WebXACT (also referred Bobby), WAET, K-WHA, A-Checker, etc. These techniques has the advantage of providing useful evaluations of web accessibility as well as quantitative results [6]. In [23] the authors found that 78.9% of the webmasters were aware that there are automatic tools to check web accessibility. This means that the missing knowledge is not the main reason for the absence of development of accessible websites. The web accessibility evaluation in e-commerce has been limited, in particular in the context of business-to-consumer (B2C) consumer. The following section discusses web accessibility in e-commerce.

2.3 Web Accessibility in E-Commerce

The significance of web accessibility standards in e-commerce has been known around the world [2, 45]. Noonan [32] investigated accessible e-commerce in Australia and recommends that e-commerce developers should consider accessibility in their web design. "As public organizations and private businesses rely more on web based technologies for online shopping, information, and service delivery they must implement strategies to ensure all users can fully access web content" [36], and proposes a web accessibility model to benefit all public organizations and private businesses. As noted in [37], "e-commerce sites lose up to 50% of potential online sales because users cannot find what they want". An accessible website provides a satisfying experience to end-users, hence increasing sales and revenue for seller [5]. In [29] the authors analysed usability and accessibility errors of African e-commerce websites compared to Europe using an automated tools. In [30] the authors recommend putting their own accessibility guidelines for African countries to ensure accessibility for all users.

According to Dolson [10], "The physical disabilities of a merchant's visitors are a factor that he or she should consider". In [40] the author believes that getting more loyal customers and avoid legal challenges are the other two reasons to design for accessibility. If the consumer gets the relevant information, then the trustworthiness of the website is increased and hence leads to higher purchase intention [14]. Faulkner [12] developed a Web Accessibility Toolbar (WAT) for Internet Explorer in collaboration with Vision Australia, the Paciello Group (Europe) and of the web accessibility tools consortium, to assist in evaluating a web page for compliance to the Web Content Accessibility Guidelines (WCAG 2.0). It is well recognized that information on the e-commerce websites varies in quality. To the extent that consumer perceive that e-commerce website presents quality information, they are more expected to have confidence and will perceive the merchant as trustworthy [21]. Therefore, there is a need of quality measurement criteria accessibility for e-commerce [16].

3   Methodology

The most practical method for measuring website accessibility is content analysis. Therefore, content analysis approach is used to investigate web accessibility in e-commerce websites. In particular, the unit of measure for this study is Australian Business-to-Consumer (B2C) e-commerce websites. The sample for this study was generated by region from Alexa, a provider of global web metrics. In (www.alexa. com) website, by clicking the link "Browse Top Sites" and selecting country Australia, 500 website were provided. After deletion of irrelevant links (such as, non B2C e-commerce websites), remaining top 30 B2C e-commerce websites was finalized for further analysis.

3.1   Instrument

Automated analyses were performed using an open source web accessibility evaluation tool called "A-Checker" version 0.8.6 [1] to test all the web pages of selected websites for conformance to web content accessibility guidelines version 2.0 (WCAG 2.0). The online web service "A-Checker" is developed by a research group at the University of Toronto [48] that tests single web page for accessibility conformance. "A-Checker" identifies three types of problems.

- **Known Problems (KP)**: These are problems that must be fixed and have been identified as accessibility barriers.
- **Likely Problems (LP)**: These are problems that are likely to be fixed and have been identified as probable barriers.
- **Potential Problems (PP)**: These are problems that require a human decision for modifying or not to modify your webpage.

3.2   Procedure

Data analysis took place in November 2015. The homepages of the selected B2C websites were tested in order to gain insight into what issues web pages might contain. The URL for each web page was entered into a required field and checked for accessibility. Options such as 'HTML validator' and 'CSS validator' were enabled and 'view by guideline' report format was selected. All three types of problems (known problems, likely problems and potential problems) were checked and recorded for each level of web content accessibility (WCAG 2.0) compliance of each website. Such as, each webpage was tested for Level A, Level AA and Level AAA of WCAG 2.0. Data (errors) were placed into an Excel sheet and descriptive statistics was conducted using SPSS v.22. If the webpage had the minimum level of conformance error (Level A), it failed the test. If no error were recorded, the webpage passed the test.

4   Results

The results showed that Australian online stores are not paying attention to at least a minimum level of conformance (Level A) of web content accessibility guidelines for their online business websites. None of the 30 Australian B2C e-commerce websites meets the minimum success criteria (Level A) of WCAG 2.0. Table 2 shows the descriptive statistics of the errors identified. For "Level A" conformance (the minimum level of conformance), a high number of known problems (KP) in Australian websites (mean = 19.1) showed accessibility barriers that must be fixed. The results also showed quite a high number of (Level A) potential problems (PP) that requires human decision to fix. The figures are worse for 'Level AA' known problems (KP) are (mean = 119.1) followed by 'Level AAA' known problems (KP) (mean = 123.6). The occurrence of these accessibility barriers will make it difficult for people with disabilities to shop online.

Table 2 Group statistics of success criteria

Success criteria	Mean
Known Problem (Level A)	**19.13**
Likely Problems (Level A)	5.87
Potential Problems (Level A)	782.13
Known Problem (Level AA)	**119.17**
Likely Problems (Level AA)	5.74
Potential Problems (Level AA)	833.48
Known Problem (Level AAA)	**123.65**
Likely Problems (Level AAA)	1.48
Potential Problems (Level AAA)	863.39

The main problems reported in Australian B2C e-commerce websites were:

- Missing label for text element and input assistance such as "label text is empty" (65% of pages).
- In addition, problems that need to be fixed are: the contrast between the color of text and its background. The most and severe violations against web content accessibility guidelines (WCAG 2.0) were:
- Level A 1.3. Ensure that information and structure can be separated from presentation (55% of pages).

  - Success Criteria 1.3.1 Info and Relationships (Level A)
  - "Input element, the type of "text", missing an associated label"

- Level A 3.3. Input Assistance: Help users avoid and correct mistakes (73% of pages).

  - Success Criteria 3.3.2 Labels or Instructions (Level A)
  - "Label text is empty"

- Level AA 1.4. Distinguishable: Make it easier for users to see and hear content including separating foreground from background (68% of pages).

  - Success Criteria 1.4.3 Contrast (Minimum) (Level AA)
  - "The contrast between the color of text and its background for the element is not sufficient to meet WCAG2.0 Level AA".

5 Discussion

The analysis reveals a growing need for addressing the current problem of web accessibility in Australian B2C e-commerce. Websites are not designed with equal access for all users in mind. Table 3 presents the complexity levels of some accessibility errors, which shows how easy it is to fix the errors [13, 46]. It is highly desirable that e-commerce firms make greater efforts to ensure that the consumers with disabilities have equal access to their websites.

The Australian government has endorsed the Web Content Accessibility Guidelines (WCAG 2.0) and made a mandatory requirement for all e-government websites to conform to Level AA [33]. The Australian government also needs to legitimate and strongly encourage e-businesses to develop accessible e-commerce applications for the widest audience. In a worldwide context, web development is now growing for e-business. From a human computer interaction viewpoint, accessible websites are becoming ever more important. The web technology creates new opportunities for e-commerce firms, but as well as challenges. Companies invest in e-business since the web has become the platform to perform business efficiently and effectively. Many business models are applied to attract and engage consumers to revisit their websites frequently. However, the presentation of B2C web design features are not conveyed through web accessibility guidelines to

Table 3 Complexity level of some web accessibility errors

Type of accessibility errors	Priority (Level)	Ease of fixing
Alt text is not used for each region of an image map	1 (Level A)	Easy
For tables not used for layout (e.g., spreadsheet), identify headers for table rows and columns	1 (Level A)	Easy
If color is used to convey information, ensure information is also provided in another way	1 (Level A)	Moderate
Did not provide alt text for images that convey content	1 (Level A)	Easy
Did not provide label tags for form fields	1 (Level A)	Moderate
Page does not have logical heading structure	1 (Level A)	Moderate
Programmatic objects should not cause screen to flicker	1 (Level A)	Hard
Used tables to format text documents in columns	2 (Level AA)	Hard
Did not ensure that background and foreground colors contrast sufficiently	2 (Level A)	Easy
Did not provide descriptive titles for links	2 (Level AA)	Easy
Used absolute (pixels) rather than relative sizing and positioning (% values)	2 (Level AA)	Moderate
When scripts created pop-up windows or changed the active window, page did not ensure that user was aware that this was happening	2 (Level AA)	Moderate
Used deprecated language features	2 (Level AA)	Moderate
Did not identify language of text	3 (Level AAA)	Moderate
Did not specify logical tab order among form controls, links, and objects	3 (Level AAA)	Moderate
Did not provide keyboard shortcuts to frequently used links	3 (Level AAA)	Moderate
Did not provide summary and caption for tables	3 (Level AAA)	Moderate
Did not group related links	3 (Level AAA)	Moderate
Did not provide linear text alternative for tables that laid out content in parallel word-wrapped columns	3 (Level AAA)	Hard
Did not provide abbreviations for long row or column labels	3 (Level AAA)	Easy
Live regions are not specified with appropriate WAI-ARIA attributes	3 (Level AAA)	Hard

consumers of all ages including disabilities. Therefore, e-commerce websites must also follow web accessibility by law [42].

5.1 Contributions and Recommendations

It is extremely recommended to B2C e-commerce firms to ensure that their website is according to WCAG 2.0 [14], which means that consumers with disabilities can purchase online. In addition, based on the [7, 46] recommendations, this study present the following suggestions to B2C e-commerce websites for people with disabilities such as sensory (hearing and vision), motor (limited use of hands) and cognitive (language and learning disabilities) impairments.

Consumer with vision difficulties: Web accessibility is particularly important since blind or color blind consumer has much more difficulty browsing the web [4]. Therefore,

- Product images should be displayed denoting their purpose and not appearance. Such as the use of ALT-tags to allow screen reader to skip unimportant images.
- Use short description for images. For-example, by stating a clour name.
- "People with low vision can use screen readers such as Job Access with Speech (JAWS), available from http://www.freedomscientific.com/products/fs/jaws-product-page.asp or Window Eyes (i.e., a screen reader for Microsoft Windows)".
- "In order to learn how a screen reader for a person with vision loss would orally present the text of a website, developers can use a Firefox plug-ins such as Fangs Screen Reader Emulator (https://addons.mozilla.org/en-US/firefox/addon/fangs-screen-reader-emulator/)" [46].
- Color Oracle software can be used by designers freely (from http://colororacle.org) for color-blind people [21].
- Avoid text font that are difficult to read with limited resolutions. The Color Blindness Simulator is also available online (http://www.colblindor.com/coblis-color-blindness-simulator/) to close this gap.
- Avoid low contrast design. Achieving effective contrast is easy by following the WCAG 2.0 recommendations, such as effective colour contrast ratio. There are amply of tools out there calculate contrast ratio. Such as Lea Verou's Contrast Ratio checker. http://leaverou.github.io/contrast-ratio/
- Coady [54] provided five ways to improve e-commerce design for color-blind users.

Consumer with hearing difficulties

- Such consumer should be provided with text captioning (closed captioning) for all audio content.
- Variety of tools available of closed captioning such as MAGpie 2, Docsoft software and YouTube also offer closed captioning services.

Consumer with learning and language difficulties

- Text on web pages should also be resizable to improve readability.
- Allow the consumer enough time when requiring input, such as in making online transactions.

The results of this study may help online shopping managers who could use the insights analysed in this research to modify their approach. Developers and website designers can use these understandings to increase desirable outcomes by focusing the web content accessibility guidelines (WCAG 2.0), to increase the chances for an online business to succeed in countries with diverse degrees of Internet users. Practical implications extend to business firms to make changes to their online business strategies to trigger their online sale better by targeting consumers with disabilities.

5.2 Conclusion and Limitations

The results show that B2C e-commerce websites in Australia are not paying attention to meet at least the minimum success criteria (Level A) of web content accessibility guidelines (WCAG 2.0). With the widespread of mobile technology, online shopping has grown significantly in recent years. Therefore, consumers with disabilities are increasingly demanding an accessible online shopping. Web content accessibility gives the opportunity for disabled people to use websites. Web accessibility for B2C websites is also important for legal and a business reasons point of view. It is also helpful to increase serviceability of B2C to engage online consumers of all ages and to increase e-retailer reputation and revenue.

This study has limitations, the B2C websites selected, which may affect the generalization of the study to other specific B2C websites. In addition, other online accessibility evaluation tools and expert evaluation may report diverse web accessibility errors.

References

1. A-Checker. Web accessibility checker. http://achecker.ca/checker/index.php (2006). Accessed 15 Sept 2015
2. Bernard, E.K., Makienko, I.: The effects of information privacy and online shopping experience in e-commerce. Acad. Mark. Stud. J. **15**, 97–112 (2011)
3. Brajnik, G., Yesilada, Y., Harper, S.: The expertise effect on web accessibility evaluation methods. Hum. Comput. Inter. **26**(3), 246–283 (2011)
4. Brunsman-Johnson, C., Narayanan, S., Shebilske, W., Alakke, G., Narakesari, S.: Modeling web-based information seeking by users who are blind. Disabil. Rehabil. Assistive Techn. **6**(6), 511–525 (2011)
5. Chelule, E.: E-commerce usability: do we need guidelines for emerging economics? In: IADIS International Interfaces and Human Computer Interaction 2010 (2010)
6. Cho, J., Lee, D., Hong, S.: A study on the web site accessibility evaluation. Bus Res. Cent. Dong-A Univ **26**(26), 161–179 (2005)
7. Crow, K.L.: Four types of disabilities: their impact on online learning. Techtrends **52**(1), 51–55 (2008)
8. Cyr, D.: Website design, trust and culture: an eight country investigation. Electron. Commer. Res. Appl. (0), 1–12 (2013)
9. Cyr, D., Bonanni, C., Bowes, J., Ilsever, J.: Beyond trust: web site design preferences across cultures. J. Glob. Inf. Manag. (JGIM) **13**(4), 30 (2005)
10. Dolson, J.C.: Accessibility: how many disabled web users are there? http://www. practicalecommerce.com/articles/1417-Accessibility-How-Many-Disabled-Web-Users-Are-There (2009). Accessed 10 Oct 2015

11. Éthier, J., Hadaya, P., Talbot, J., Cadieux, J.: Interface design and emotions experienced on B2C Web sites: empirical testing of a research model. Comput. Hum. Behav. **24**, 2771–2791 (2008)
12. Faulkner.: S. Web Accessibility Toolbar for IE. http://www.visionaustralia.org/business-and-professionals/digital-access/resources/tools-to-download/web-accessibility-toolbar-for-ie—2012 (2012). Accessed 15 Sept 2015
13. Flowers, C., Bray, M., Algozzine, R.: Content accessibility of community college websites. Community Coll. J. Res. Pract. **25**(7), 475–485 (2001)
14. Ganguly, B., Dash, S.B., Cyr, D.: The effects of website design on purchase intentionin online shopping: the mediating role of trustand the moderating role of culture. Int. J. Electron. Bus. **8**(4/5), 302–329 (2010)
15. Golden, N.: Access this: why institutions of higher education must provide access to the internet to students with disabilities. J. Entertainment Technol. Law, **10**(3), 363–411 (2008)
16. Hasan, L., Abuelrub, E.: Assessing the quality of web sites. Appl. Comput. Inform. **9**(1), 11–29 (2011)
17. Hong, S., Cho, J., Lee, D.: Government website accessibility: comparison between Korea and the United States. Inf. Syst. Rev. **7**(1), 81–96 (2004)
18. Hyun, J., Hong, K., Shin, K., Min, H.: Web accessibility compliance of major web sites in Korea. J. Rehabil. Eng. Assist. Technol. Soc. Korea **1**(1), 37–43 (2007)
19. Hyun, J., Kim, B.: Web accessibility compliance of internet bankings in Korea. J. Korea Soc. Inf. Technol. **7**(2), 77–93 (2008)
20. Kang, Y., Hong, S., Lee, H., Cha, Y.: Website accessibility evaluation of the welfare centers for the disabled. J. Korea Acad-Ind. Coop. Soc. **12**(11), 5260–5271 (2011)
21. Kim, D.J., Ferrin, D.L., Rao, H.R.: A trust-based consumer decision-making model in electronic commerce: the role of trust, perceived risk, and their antecedents. Decis. Support Syst. **44**(2), 544–564 (2008)
22. Kuzma, J.M.: Accessibility design issues with UK e-government sites. Gov. Inf. Q. **27**, 141–146 (2010)
23. Lazar, J., Alfreda, D., Greenidge, K.: Improving web accessibility: a study of webmaster perceptions. Comput. Hum. Behav. **20**, 269–288 (2004)
24. Lazar, J., Sears, A.: Design of e-business web sites. In: Handbook of Human Factors and Ergonomics, pp. 1344–1363. John Wiley & Sons, Inc. (2006)
25. Lee, J., Lee, B.: Web accessibility evaluation of cyber universities' contents in Korea. J. Korea Contents Soc. **7**(4), 224–233 (2007)
26. Li, S.-H., Yen, D.C., Lu, W.-H., Lin, T.-L.: Migrating from WCAG 1.0 to WCAG 2.0—A comparative study based on web content accessibility guidelines in Taiwan. Comput. Hum. Behav. **28**(1), 87–96 (2012)
27. Loiacono, E.: web accessibility and corporate America. Commun. ACM **47**(12), 82–87 (2004)
28. Luchtenberg, M., C. Kuhli-Hattenbach, Y. Sinangin, C. Ohrloff, R. Schalnus.: Accessibility of health information on the internet to the visually impaired user. Ophthalmologica. **222**(3), 187–193 (2008)
29. Maswera, T., R. Dawson, J. Edwards.: Analysis of usability and accessibility errors of e-commerce websites of tourist organisations in four african countries. In: Frew, A. (ed.) Information and Communication Technologies in Tourism 2005, pp. 531–542. Springer, Vienna (2005)
30. Maswera, T., Edwards, J., Dawson, R.: Recommendations for e-commerce systems in the tourism industry of sub-Saharan Africa. Telematics. Inform. **26**(1), 12–19 (2009)
31. Morgon, R.: State of the nation's $24billion online retail trade: internet shopping becomes the new Australian norm. http://www.abc.net.au/news/2013-06-04/more-than-50-per-cent-of-australians-shopping-online/4731590 (2013). Accessed 10 Sept 2015

32. Noonan, T.: Accessible e-commerce in Australia. http://www.timnoonan.com.au/ecrep10. htm_Toc463495402 (1999). Accessed 10 Sept 2015
33. NTS. Accessibility. http://webguide.gov.au/accessibility-usability/accessibility/ (2011). Accessed 11 Sept 2015
34. O'Grady, L.: Accessibility compliance rates of consumer-oriented Canadian health care web sites. Inform. Health Soc. Care 30(4), 287–295 (2006)
35. Oh, K., Kim, Y.: Investigation of web accessibility on hospital websites in Korea. Proc. Korean Soc. Internet Inf. 9(2), 375–380 (2008)
36. Ou, C.X., Sia, C.L.: Consumer trust and distrust: An issue of website design. Int. J. Hum. Comput. Stud. 68(12), 913–934 (2010)
37. Park, H., Ban, C.: Implementation and evaluation of the on-line aptitude test system for people with visual impairment supporting web accessibility. Disabil Employ 20(1), 51–78 (2010)
38. Potter, A.: Accessibility of Alabama government web sites. J. Gov. Inf. 29, 303–317 (2000)
39. PwC. Australia and New Zealand online shopping market and digital insights. http://www.pwc.com.au/industry/retail-consumer/assets/Digital-Media-Online-Shopping-Jul12.pdf (2012). Accessed 15 Sept 2015
40. Roggio, A.: The importance of web accessibility for e-commerce. http://www.getelastic.com/the-importance-of-web-accessibility-for-ecommerce/ (2008)
41. Seo, E., Kim, H.: Comparative analysis of web accessibility in national libraries. J. Korean Soc. Libr. Inf. Sci. 42(3), 345–364 (2008)
42. Smallman, W.: Should e-commerce websites support web accessibility by law? http://www.blahblahtech.com/2006/11/should-e-commerce-websites-support-web-accessibility-by-law.html (2006). Accessed 10 Sept 2015
43. Sohaib, O., K. Kang.: The importance of web aceesibility in business to-consumer (B2C) Websites'. In: 22nd Australasian Software Engineering Conference (ASWEC 2013), pp. 1–11 (2013)
44. Sohaib, O., Kang, K.: Individual level culture influence on online consumer iTrust aspects towards purchase intention across cultures: A S-O-R model. Int. J. Electron. Bus. 12(2), 141–162 (2015)
45. Sohaib, O., K. Kang.: The role of technology, human and social networks in serviceable cross-cultural B2C websites. In: 19th International Business Information Management Conference (IBIMA), Barcelona, Italy (2012)
46. Solovieva, T.I., Bock, M.J.: Monitoring for accessibility and university websites: meeting the needs of people with disabilities. J. Postsecondary Educ. Disabil. 27(2), 113–127 (2014)
47. Sullivan, T., R. Matson.: Barriers to use: usability and content accessibility on the web's most popularsites. In: Proceedings of the 2000 Conference on Universal Usability (2000)
48. W3C. Complete list of web accessibility evaluation tools http://www.w3.org/WAI/ER/tools/complete (2012). Accessed 5 Nov 2015
49. W3C. Introduction to web accessibility. http://www.w3.org/WAI/intro/accessibility.php (2005)
50. W3C. Web content accessibility guidelines (WCAG) 2.0. http://www.w3.org/TR/WCAG20/ (2008). Accessed 1 Oct 2016
51. Kamoun, F., Almourad, M.B.: Accessibility as an integral factor in e-government web site evaluation. Info. Technol. & People, 27(2), 208–228 (2014)
52. King, A., Evans, G., Blenkhorn P.: Webbie: a browser for visually impaired people. In: Proceedings of the 2nd CambridgeWorkshop on Universal Access and Assistive Technology, pp. 35–44 (2004)
53. Power, C., Jurgensen, H.: Accessible presentation of information for people with visual disabilities. Univ. Access. Inf. Soc. 9(2), 97–119 (2010)

54. Coady, C.: 5 Ways to improve your ecommerce design for colourblind users. https://www.shopify.com.au/partners/blog/86314118-5-ways-to-improve-your-ecommerce-design-for-colourblind-users (2016). Accessed 5 Oct 2016
55. WHO.: New world report shows more than 1 billion people with disabilities face substantial barriers in their daily lives. www.who.int/mediacentre/news/releases/2011/disabilities_20110609/en/index.html (2011). Accessed 14 Sept 2016
56. Australian Bureau of Statistics (ABS).: Disability, ageing and carers, Australia: first results. http://www.abs.gov.au/ausstats/abs@.nsf/mf/4430.0.10.001 (2015). Accessed 5 Oct 2016

Endogenously Emergent Information Systems

J. Iivari

1 Introduction

Information systems are growing more complex and autonomous systems of systems. Salvaneschi [70] describes it nicely: "Large information systems are composed of dozens of software applications—programs that typically implement a business process or part of it. Applications may be developed in house or acquired from vendors and possibly adapted. During the evolution the information system grows, integrating more and more applications and changing the existing ones. The evolution is managed by different vendors and development teams working only on parts of the whole system" (pp. 8–9).

IBM has suggested four features of autonomic systems: self-configuring, self-optimizing, self-healing and self-protecting [40]. Nielsen et al. [63] characterize systems of systems with autonomy of the constituent systems, their operational independence but interdependence within the whole, distribution, evolution, dynamic reconfiguration, interoperability and emergence. The focus of this paper lies in "emergence".

Information systems have been characterized as "emergent" by a number of researchers. The term "emergent" and "emergence" are ambiguous, however, with a number of interpretations and meanings. These concepts have been of considerable interest in Computer Science, but not so much in the IS literature.

So, the purpose of this paper is to analyze and clarify the concepts of "emergence" in the context of information systems and to discuss their implications to IS

A prior version of this paper has been published in the ISD2016 Proceedings (http://aisel.aisnet.org/isd2014/proceedings2016).

J. Iivari (✉)
University of Oulu, Oulu, Finland
e-mail: juhani.iivari@oulu.fi

© Springer International Publishing Switzerland 2017
J. Gołuchowski et al. (eds.), *Complexity in Information Systems Development*,
Lecture Notes in Information Systems and Organisation 22,
DOI 10.1007/978-3-319-52593-8_7

research. The paper pays special attention to endogenous dynamic emergence of information systems, implying that "emergence" is due to the complexity of the system and its operational interaction with its environment. There are three reasons for this focus. Firstly, as noted above, information systems have grown more complex, often being systems of systems [70] or at close to them [32]. This complexity makes them prone to emergent behavior. Secondly, some of this emergent behavior may be undesirable as illustrated for example by the unintended sudden acceleration of cars due to software and the anomalous stock market behaviors [74, 92]. Thirdly, existing IS research has largely omitted this endogenous emergence. The reason for this neglect may be that the IS community has not had a special concept to make the phenomenon explicit.

2 The Concept of Emergence

According to [58] "emergence refers the phenomenon whereby the macroscopic properties of a system arise from the microscopic properties (interactions, relationships, structures and behaviours) of its constituents" (p. 422). It has been widely discussed in biology, psychology, physics, systems theory, philosophy and so on [20, 21, 71, 72]. Quite interestingly, it has also been of considerable interest in Computer Science and in particular in Artificial Intelligence (e.g. [1, 2, 37, 60, 84]) but not so much in the IS literature, [39, 46, 58, 61] as exceptions.

"Emergence" continues to be a contested concept and it is difficult, if not impossible, to provide a definition that would be accepted by all. It is often characterized by phrases such as "the whole is more than the sum of its parts", "much coming from little", "coming into being". So, this paper does not attempt to provide any definite definition, but conceptualizes "emergence" in terms of a number of characteristics shared by "emergent" systems. However, in the case of "emergence" I will focus on the "dynamic emergence" rather than on the "static emergence" [1, 2]. Static emergence can be illustrated by emergent properties such as the number of bedrooms in a house or the durability of a spider web. Dynamic emergent properties change over time. They represent emergent behavior as "coming into being" [34].

Dynamic emergent behavior may be designed and in that way anticipated (= functionality) or non-designed and unanticipated. Normally, the designed emergent behavior is desirable, whereas non-designed, unanticipated behavior may be either desirable or undesirable. Figure 1 introduces the resultant classification of emergent properties, inspired by Ferreira et al. [28].

One way to open the concept "emergence" is to look at characteristics of phenomena that are considered emergent: complexity of the system, its interaction with the environment, learning and adaptation, lack of central control, and unpredictability.

Emergence is often associated with complex systems [37] and especially complex dynamic systems [8]. Bar-Yam [8] notes that a complex system of interdependent parts may exhibit complex emergent behavior. There are also explicit attempts in Computer Science to apply the ideas of complex adaptive systems to

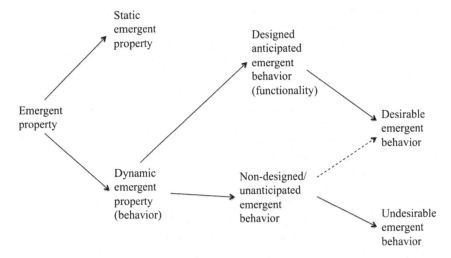

Fig. 1 A classification of emergent properties

develop models of emergent computation that are explicitly based on "emergence" [60]. Emergent behavior or functionality in these computational models cannot be reduced to the behaviors of the agents the system is composed of [29].

Complex dynamic systems are usually open systems. Wegner et al. [90, 91] point out the traditional algorithmic model of computing based on the Turing machine is limited, since Turing machines cannot accept external input while computing.[1] As an alternative [90] proposes "interactive computing", which allows interactions with the environment while computing. If the system simultaneously interacts with numerous environmental objects, it obviously increases the behavioral complexity of the system. So, one can anticipate that not only the internal complexity of the system in terms of the number of its elements and their interdependencies but also the interaction complexity with the environment may produce emergent behavior.

Holland [37] claims that "Any serious study of emergence must confront learning" (p. 53). He illustrates how an adaptive system governed by relatively few and simple rules can exhibit emergent behavior. Although machine learning has a long tradition in Artificial Intelligence, learning information systems have not formed a notable topic in mainstream IS research.

Holland [37] also associates emergence with lack of central control. One can analyze the existence of centralized control in the operational system—whether there is a subsystem that centrally controls the whole system—and existence of centralized control especially when an information system are developed in a

---

[1]Note that there seems to be some differences of opinion of what Turing had in mind or what Turing Machine as a model of computing implies [17].

distributed way. Referring to [63] one can argue that the trend is towards systems without centralized control.

Predictability has been discussed especially in the context of reducibility of emergent phenomena, i.e. to what extent they (at the system level) can be explained in terms of the lower-level mechanisms (e.g. system components and their interactions). The question has been of interest especially to philosophers [15, 16, 47, 48]. Chalmers [15] distinguishes strong emergence and weak emergence: Strong emergence assumes that an emergent phenomenon arises from the low-level domain, but truths concerning that emergent phenomenon are not deducible even in principle from truths in the lower-level domain. Weak emergence assumes that truths concerning the phenomenon just are unexpected given the principles governing the low-level domain. Kim [47] distinguishes inductive predictability and theoretical predictability of emergent phenomena. He contends that one can inductively predict emergent phenomena: Having observed that an emergent phenomenon E occurs whenever any system has a specific low-level state S, one may predict that a particular system will have the emergent phenomenon E at specific time, if one knows that low-level state of the system will be S at the specific time. Inductive predictability does not imply theoretical predictability so that even full information of the state of the low level domain would not allow prediction of the emergent phenomenon. One objection to the theoretical predictability is that the emergent phenomenon E is not a concept belonging the low-level domain.

In line with [37] the present paper adopts a pragmatic rather than a philosophical position to the predictability of emergent phenomena, focusing especially on information systems. Since the normal functionality of a system is expected emergent behavior of the system, I will focus on the unpredictable emergent behavior that is unexpected, coming as a surprise. Complex software and information systems tend to have such unexpected emergence, since the prediction of their behavior is usually difficult, due to the fact that the state space of the system is very large, the system rarely returns to a state already visited, especially if the system is able to learn and adapt [37, 67].[2]

3   The Concept of Emergence in the Mainstream IS Literature

Since in the ordinary language the word "emerge" may be used as a synonym to words such as "appear" and "rise", it is quite difficult to conduct a bibliographic search on the more "technical" use of the word in the context of information

---

[2]Predictability is, of course, a matter of degree. Therefore, many qualities of information and software systems such as reliability, maintainability, efficiency can be regarded as emergent properties—they are system-level qualities hard to reduce to the system components and usually not completely predictable.

systems. Based on the authors' familiarity with Information Systems, I tried a forward search [89] in which I used as additional keywords the names of the authors of the early articles on emergence in the context of information systems [36, 53, 57, 65, 78–80], limiting the search to the time after the article was published.

Bibliographic searches, focusing on journal articles, using these lists identified a number of additional articles that refer to "emergence" in the spirit of the previous section (see Table 1). The list is not necessarily exhaustive, but likely representative and indicates continued interest in "emergence" in the context of information systems.

Table 1 summarizes the findings, distinguishing organization, IS development (ISD) process (incl. design and implementation), IS use, and IS artifact as phenomena which may considered "emergent". Table 1 also shows that much of the reviewed literature emphasizes emergence in the context of organizations. Although this literature usually does not explicate the micro phenomenon (agents) that generates the emergence, one can easily construe that an organization is a continuous emergent achievement of its members, stakeholders and other organizations it interacts with.

It is also common to characterize the ISD process as emergent. Although also this stream is not very explicit on the micro phenomena that give rise to emergence, one can imagine that it is an outcome of negotiations between users, managers, designers, vendors and other stakeholders during the ISD process. Furthermore, the ISD design process has become more distributed both organizationally and temporally without (complete) centralized control (see Sect. 4.5). This distribution has made the ISD process and its outcome emergent. Overall, this emergence has resemblance with the ideas of "emergent design" (design as a verb) [14, 25], but contrary to these examples this paper underlines the difficulty of centralized control in this context.

There is also research that considers IS use emergent. Among this stream Nan [61] is conceptualizes the emergence to rise from the interaction between user, technology and task and the interaction between users.

Table 1 also shows that a number of references have recognized the IS artifact as emergent. Yet, it has mostly been quite implicit: the emergent ISD process is assumed to lead to emergent IS artifact so that one can speak about "emergent design" (design as a noun), "emergent requirements", "emergent architecture", "emergent structure", analogously to emergent strategy in [59]. In this case "emergence" is assumed to be an outcome of an emergent development process.

So, the mainstream IS literature cited in Table 1 mostly assumes emergence to be an outcome of exogenous, although, complex design agency. I suggest that "emergence" of IS artifacts may also be more endogenous, inherent to them, resulting from the internal interaction of the subsystems and their interactions with the (dynamic) environment. The purpose of the following section is to have a more detailed look at endogenously emergent behavior of information systems.

Table 1 Primary focus on "emergent" phenomena in the context of information systems

Article	Organization	ISD process	IS use	IS artifact
Markus and Robey [57]	x		x	
Truex and Klein [80]	x	x		x
Lyytinen and Ngwenyama [53]	x	x	x	x
Hirschheim et al. [36]	x	x		x
Iivari and Hirschheim [42]	x	x		x
Orlikowski [65]	x		x	
Ngwenyama [62]	x	x	x	x
Karsten [44]	x	x	x	x
Lycett and Paul [52]	x	x		x
Truex et al. [78]	x	x		x
Truex et al. [79]	x	x		x
Orlikowski [66]		x	x	x
Baskerville and Siponen [7]	x			
Bergman et al. [9]		x		x
Markus et al. [56]	x	x		x
Thompson [76]			x	
Levina [50]		x		x
Luna-Reyes et al. [51]		x		
Allen and Varga [4]	x			
Constantinides and Barrett [18]	x	x	x	x
Corea [19]	x		x	
Curseu [22]	x			
Dreyfus and Iyer [24]				x
Bjørn and Ngwenyama [10]	x			
Patel et al. [68]	x	x		x
Wagner et al. [87]	x	x	x	x
Baker [6]	x	x		
Holmström and Sawyer [38]	x			x
Nan [61]			x	
Essen and Lindblad [26]		x		x

4 Towards Endogenously Emergent Information Systems

4.1 Introduction

Contrary to the IS literature, the issue of endogenously emergent systems has been extensively addressed in biologically inspired Computer Science and Software Engineering (e.g. [11, 29, 40, 55, 63, 83]), in particular. The purpose of this section is to discuss endogenous emergent behavior from the viewpoint of information systems.

This paper interprets an "information system" (IS) to be a computer-based system whose purpose "is to supply its groups of users (…) with information about a set of topics to support their activities" [35].[3] In more morphological terms, an "information system" is a combination of application software and digital information content [86]. According to this interpretation an information system is specific to the organizational (or inter-organizational) context in which it is implemented and that pure software (such as an ERP package) is not an information system.

Following Carvalho [12], we can conceive an information system as a set of interrelated active objects that deal with symbolic objects (information) and whose agents are computers or computer-based devices. Each active object includes a piece of information (or more strictly data embedding or conveying information) and has a number of operations to access data from the environment, to display data, to process data, and to communicate with other active objects of the system. The active objects may either be transient objects or more persistent database objects, storing structured data, electronic documents, websites, knowledge repositories, for example. The granularity of symbolic objects (information content) may vary from simple factual statements to long unstructured documents. In principle, each symbolic object (e.g. each factual statement) may have its own active object.

The definition of an information system is significant, since it naturally has a huge impact on what is endogenous and exogenous. In my view, users and organization belong to the environment of an information system as well as its designers (developers), and an information system is the artifact to be designed and to be used (called "IS artifact" below for brevity).

We recognized internal complexity, dynamic interaction with the environment, learning and adaptation, lack of central control, and unpredictability as typical characteristics of emergent systems. In the following I will argue that modern and especially future IS artifacts share many of these features and are emergent in that sense.

4.2 Internal Complexity

IS artifacts are increasingly complex. The internal complexity of IS artifacts opens the door for emergence as in the case of any complex systems. Furthermore, an IS artifact may be so complex that nobody—a single individual or designer group collectively—understands its totality [82]. As a consequence the system may include or develop unanticipated emergent features. This possibility of emergent features poses considerable reliability, safety and privacy challenges in the case of

---

[3]I interpret "organization" in a broad meaning here so that in addition to formal organizations it covers more informal organizations such as families and various online communities.

many information and software systems (e.g. [27, 73, 88]). Emergent features as security risks imply that the risk is not because of a local bug, but a result of complex interaction of the component systems, interaction that is extremely difficult to figure out during the design because of the complexity of the system and the question is about run-time interaction.

One should also note that the configuration of an IS artifact in terms of its active objects may be dynamic. In massive, large-scale, wide-area computing networks (such as Internet) and mobile ad hoc networks nodes (active objects) may join and leave the network and connection between nodes may fail [5]. These configurational dynamics naturally increases the complexity of the system and chances of unanticipated emergent behavior.

Such complex IS artifacts may also comprise massive amount of potentially heterogeneous information content (data). Tolk et al. [77] distinguish six levels of interoperability systems. In addition to the technical interoperability (a communication protocol for exchanging data between the systems) and syntactic interoperability (a common structure to exchange data), they identify semantic interoperability (a shared meaning of data, i.e. its information content), pragmatic interoperability (awareness of methods and procedures applied in each subsystem), dynamic interoperability (awareness of the state changes in assumptions and constraints each subsystem implies), and conceptual interoperability (assumptions and constraints of the abstraction of reality).[4] Although the subsystems or components are interoperable at the technical and syntactical level so that the subsystems are able to communicate with each other, it does not assure that the system as a whole exhibits the desired emergent functionality and avoids harmful behavior.

Potentially, this massive information content includes hidden patterns to be discovered. All research on "big data" and data mining rests on this potentiality. These hidden patterns can be considered as static emergent properties, which arise from the individual data and their relationships.

4.3   Interaction with the Environment

IS artifacts have not only grown internally complex, but their interaction with the environment has also become more complex, Internet-of-things as the latest trend. Valckenaers et al. [82] divides problems into one-shot problems and going concerns, claiming that real-life problems are mostly of the latter type. One-shot computational problems can be solved using algorithmic models of computing, but going concerns require interactive computing [90]. Although one should not confuse interactive computing, distributed computing and parallel computing,

---

[4]These levels are developed keeping simulation models in mind, but illustrate that deliberate and safe integration of systems requires getting acquainted with a huge amount meta-information the systems to be integrated.

obviously interactive computing that is distributed and includes parallelism forms a situation where emergence as the system's unanticipated features and behavior is most likely.

Goldin et al. [30] discuss database-oriented IS artifacts emphasizing that their dynamics can either be algorithmic, sequential interactive, or concurrent interactive —represented by a Turing machine, Sequential (single-stream) Interaction Machine (SIM) or Multi-stream Interaction Machine (MIM) respectively. They note that MIM provides a generic model for the IS dynamics, implying transduction of one or more autonomous input streams from external users into output streams of system feedback, accompanied by an evolving system state.[5] They also claim that the complexity of modeling an IS artifact comes from the complex nature of MIM solution spaces.

When IT becomes more mobile, pervasive and ubiquitous [54], one can expect that also IS artifacts become increasingly context-aware so that they identify and interact possibly with numerous objects in their environments such as users and other systems and objects by sensors and effectors. Context-aware computing has been of considerable research interest during the last twenty years covering for example location-awareness, environment-awareness, artifact-awareness, activity-awareness, participant-awareness, and user-awareness [31, 75].

4.4   Learning and Adaptation

Machine learning has a long tradition in Artificial Intelligence [43, 49], but to my knowledge "learning information systems" is almost a totally neglected area in mainstream IS research. On the other hand, I guess that machine learning has been discussed and applied in the contexts of special areas of IS such "intelligent information systems", "intelligent decision support systems", data mining, document/content management, and knowledge discovery, for example.[6]

The two trends—increased complexity and intensified system-environment interaction—identified above imply that IS artifacts tend to entail huge amount of information content (data) reflecting the dynamic environment. The challenge in this situation is to make sense of the meaning (semantics) of all that heterogeneous information content. It is hard to believe that any a priori defined ontology could solve the issue, but it must take place more inductively in the spirit of emergent semantics [3, 45]. For example, when a higher-level construct (e.g. a pattern in data mining) is identified, this new construct—if found useful by users (domain experts)

---

[5]Users here can be interpreted to cover not only human users in the external environment of the IS artifact, but also various objects in the environment that generate input streams and/or receive output streams.

[6]One should note that our interest here is not in initial of training of the system to classify documents, for example, but in the automatic learning by the system while in use.

—is made a part of the information content. After that users can refer to them in their interaction with the system and the system itself can structure its information content making use of the aggregated concepts.

4.5 Lack of Centralized Control

There is a clear trend towards interconnected, cooperative IS artifacts composed of independently developed application packages, software components, software agents or web services. When IS artifacts become more like systems of systems with high autonomy (managerial independence) and operational independence [63], centralized operational control of the whole system of systems becomes more unlikely. It is particularly so when the system (of systems) is automatically composed (or dynamically re-configured) or its development has been horizontally distributed without centralized control. By horizontal distribution I mean that the system is development by fairly independent teams (or individual developers) largely simultaneously, while in vertical distribution the development takes place by the same team or different teams sequentially.

Whether distributed or not, the key challenge of coordinating complex software development is the management of dependencies between software components or modules [23, 33]. Although centralized control is considered one of the "best practices" in distributed software development [13], it is difficult in practice because of, the sheer number of dependencies, which tends to explode when the size of software grows, and especially run-time dependencies are difficult or impossible to identify. As a consequence despite modular decomposition, software architecture, Application Programming Interfaces (APIs) and configuration management, software dependencies must also be coordinated by mutual adjustment requiring horizontal communication between developers and teams [13]. In light of all these challenges it is amazing that people have been able to develop such complex systems that have the desired functionality most of the time and at least not fatally harmful emergent behaviors.

4.6 Unpredictability

As mentioned above emergent behavior is to some extent unpredictable. This makes "emergence" challenging in the case of IS artifacts—how to "control" emergence so that exhibits desirable behavior and how to avoid unpredictable harmful behavior [29].

On the other hand, some unpredictability is inherently desirable in the case of IS artifacts, since one can claim that information has value only when it has surprise value.

5 Discussion and Final Comments

Although the IS literature widely refers to emergence, it has mainly focused on "emergent design" (design as a verb) of information systems and implicitly on the "emergent design" (design as a noun) of the resultant the system. It seems to me that there is an opportunity for additional research that studies conceptually and empirically different forms of distributed IS development—such as application-package-based development in the case of multiple suppliers, outsourced IS development with multiple vendors, agile development with multiple teams, free/open source development without centralized control, "re-design in use"—as instances of "emergent design" ("design" both as a verb and as a noun), recognizing that strict centralized control is problematic in these contexts.

The IS literature has largely omitted endogenously emergent behavior of information systems. In view of the fact that information systems are increasingly complex systems of autonomous systems, the question is if IS can afford to omit it and the interaction between such complex systems and different units of adopters (individuals, groups, organizations, markets, communities and societies). A noted above a big research question in the context of endogenous emergence is how "to control" it so that the system exhibits desired system-level behavior and how to avoid harmful system-level behavior, when the design focuses on agents at the micro-level [29]. The possibility of harmful emergent behavior may seriously affect individuals, organizations and society, implying that we—researchers, IS developers, politicians and the general public—should pay attention to risks of such systems [28]. So, it seems to me that there is a clear research opportunity to investigate what the ideas of systems of systems [63] and autonomic computing [40] mean to the above adopting units and stakeholders and to Information Systems as a discipline.

Contrary to Information Systems, Computer Science and Software Engineering have paid considerable attention to endogenous emergence. The question is if Information Systems could make any meaningful contribution to that discourse and open new research perspectives and directions. Overall, I have an impression that the existing literature does not address very explicitly the role of information content as a source of endogenous emergence of information systems.

When reading the literature I also encountered terms "design for emergence" and "design by emergence" [69, 81, 85]. The space does not allow me to discuss these concepts in length, but I would suggest that "design for emergence" refers to design that focuses on designing conditions and constraints to affect exogenous emergence ("emergent design") and endogenous emergence.

"Design by emergence" in my vocabulary is a design process that makes use on endogenous dynamic emergence. If we interpret this emergence as unpredictable, the combination of design and emergence seems a misnormer. However, such emergence may effectively support innovation [64]. In the case of IS artifacts, if the purpose of the system is not just to inform the users but to affect in real-time an ongoing "real" process in its environment, unpredictable emergent behavior may be disastrous. If there is a human being in between interpreting the information, he or she may observe if there is something wrong in the piece of information provided by the system, but not always. However, if the purpose of the system is to explore possibilities, unpredictable endogenous emergent system behavior may be very informative. However, "design by emergence" seems the most promising in computer game design [69, 85] and especially when designing digital fantasizing applications [41]. Design by emergence in their contexts might imply that you cannot play the same game twice or you cannot enter the same fantasy world twice, not only because the constellation of co-participants in the game instance may be different (as in multiplayer games) or the context and physical space of the game may be unique (as in location-based games), but because the game as itself is designed to comprise endogenously emergent functionality.

References

1. Abbot, R.: Emergence explained: abstractions. Complexity **12**(1), 13–26 (2006)
2. Abbot, R.: Putting complex systems to work. Complexity **13**(2), 30–49 (2007)
3. Aberer, K., et al.: Emergent semantics systems. In Semantics of a Networked World. Semantics for Grid Databases. LNCS, vol. 3226, pp. 14–43. Springer (2004)
4. Allen, P.M., Varga, L.: A co-evolutionary complex systems perspective on information systems. J. Inf. Technol. **21**, 229–238 (2006)
5. Babaoglu, O., et al.: Design patterns from biology for distributed computing. ACM Trans. Auton. Adapt. Syst. **1**(1), 26–66 (2006)
6. Baker, E.W.: Why situational method engineering is useful to information systems development. Inf. Syst. J. **21**, 155–174 (2011)
7. Baskerville, R., Siponen, M.: An information security meta-policy for emergent organizations. Logistics Inf. Manag. **15**(5/6), 337–346 (2002)
8. Bar-Yam, Y.: Dynamics of Complex Systems. Addison-Wesley, Reading (1997)
9. Bergman, M., King, J.L., Lyytinen, K.: Large-scale requirements analysis revisited: the need for understanding the political ecology of requirements engineering. Requirements Eng. **7**, 152–172 (2002)
10. Bjørn, P., Ngwenyama, O.: Virtual team collaboration: building shared meaning, resolving breakdowns and creating translucence. Inf. Syst. J. **9**, 227–253 (2009)
11. Brunner, K.A.: What's emergent in emergent computing? In: EMCSR 2002 Conference: 16th European Meeting on Cybernetics and Systems Research, vol. 1, pp. 189–192 (2002)
12. Carvalho, J.A.: Information system? Which one do you mean? In: Falkenberg, E., Lyytinen, K., Verrijn-Stuart, A. (eds.) Information Systems Concepts: An Integrated Discipline Emerging. Kluwer, pp. 259–282 (2000)
13. Cataldo, M., Bass, M., Herbsleb, J.D., Bass, L.: On coordination mechanisms in global software development. In: International Conference on Global Software Engineering (ICGSE 2007). IEEE, pp. 72–80 (2007)
14. Cavallo, D.: Emergent design and learning environments: building on indigenous knowledge. IBM Syst. J. **39**(3&4), 788–821 (2000)
15. Chalmers, D.J.: Strong and weak emergence. In: Clayton, P., Davies, P. (eds.) The Re-emergence of Emergence: The Emergentist Hypothesis from Science to Religion, pp. 1–31. Oxford University Press, Oxford (2006)
16. Clayton, P.: Conceptual foundations of emergence theory. In: Clayton, P., Davies, P. (eds.) The Re-emergence of Emergence: The Emergentist Hypothesis from Science to Religion, pp. 244–254. Oxford University Press, Oxford (2006)

17. Cockshott, P., Michaelson, G.: Are there new models of computation? Reply to Wegner and Eberbach. Comput. J. **50**(2), 232–247 (2007)
18. Constantinides, P., Michael Barrett, M.: Negotiating ICT development and use: the case of a telemedicine system in the healthcare region of Crete. Inf. Organ. **16**, 27–55 (2006)
19. Corea, S.: Mounting effective IT based customer service operations under emergent conditions: deconstructing myth as a basis of understanding. Inf. Organ. **16**, 109–142 (2006)
20. Corning, P.A.: The re-emergence of "emergence": A venerable concept in search of a theory. Complexity **7**(6), 18–30 (2002)
21. Corning, P.A.: The re-emergence of emergence, and the causal role of synergy in emergent evolution. Synthese **185**, 295–317 (2012)
22. Curseu, P.L.: Emergent states in virtual teams: a complex adaptive systems perspective. J. Inf. Technol. **21**, 249–261 (2006)
23. de Souza, C.R.B.: On the Relationships between Software Dependencies and Coordination: Field Studies and Tool Support. Ph.D. dissertation, University of California, Irvine, CA (2005)
24. Dreyfus, D., Iyer, B.: Managing architectural emergence: a conceptual model and simulation. Decis. Support Syst. **46**, 115–127 (2008)
25. Drury, M., Conboy, K., Power, K.: Obstacles to decision making in agile software development teams. J. Syst. Softw. **85**, 1239–1254 (2012)
26. Essen, A., Lindblad, S.: Innovation as emergence in healthcare: unpacking change from within. Soc. Sci. Med. **93**, 203–211 (2013)
27. Felici, M.: Capturing emerging complex interactions: Safety analysis in air traffic management. Reliab. Eng. Syst. Saf. **91**, 1482–1493 (2006)
28. Ferreira, S., Faezipour, M., Corley, H.W.: Defining and addressing the risk of undesirable emergent properties. In: Systems Conference (SysCon), pp. 836–830 (2013)
29. Gleizes, M.-P., Camps, V., George, J.-P., Capera, D.: Engineering systems which generate emergent functionalities. In: Weyns, D., Brueckner, S.A., Demazeau, Y. (eds.) Engineering Environment-Mediated Multi-Agent Systems. LNCS, vol. 5049, pp. 58–75. Springer (2008)
30. Goldin, D., Srinivasa, S., Thalheim, B.: IS = DBS + Interaction: towards principles of information system design. In: Laender, A.H.F., Liddle, S.W., Storey, V.C. (eds.) ER2000 Conference. LNCS, vol. 1920, pp. 140–153. Springer (2000)
31. Gorlenko, L., Merrick, R.: No wires attached: usability challenges in the connected mobile world. IBM Syst. J. **42**(4), 639–651 (2003)
32. Gorod, A., Sauser, B., Boardman, J.: System-of-systems engineering management: a review of modern history and a path forward. IEEE Syst. J. **2**(4), 483–499 (2008)
33. Grinter, R.: Recomposition: coordinating a web of software dependencies. Comput. Support. Coop. Work **12**, 297–327 (2003)
34. Groenewegen, P., Wagnenaar, P.: Managing emergent information systems: Towards understanding how public information systems come into being. Inf. Polity **11**, 135–148 (2006)
35. Gustafsson, M.R., Karlsson, T., Bubenko, J. Jr.: A declarative approach to conceptual information modeling. In: Olle, T.W., Sol, H.G., Verrijn-Stuart, A.A. (eds.) Information Systems Design Methodologies: A Comparative Review, pp. 93–142. North-Holland, Amsterdam (1982)
36. Hirschheim, R., Klein, H.K., Lyytinen, K.: Exploring the intellectual structures of information systems development: a social action theoretic analysis. Account. Organ. Inf. Technol. **6**(1/2), 1–64 (1996)
37. Holland, J.H.: Emergence: From Chaos to Order. Addison-Wesley (1999)
38. Holmström, J., Sawyer, S.: Requirements engineering blinders: exploring information systems developers' black-boxing of the emergent character of requirements. Eur. J. Inf. Syst. **20**(1), 34–47 (2011)
39. Hovorka, D., Germonprez, M.: Perspectives on emergence in information systems research. Commun. Assoc. Inf. Syst. **33**, 353–364 (2013)

40. Huebscher, M.C., McCann, J.A.: A survey of autonomic computing—degrees, models, and applications. ACM Comput. Surv. **40**(3), 7-1–7-28 (2008)
41. Iivari, J.: Paradigmatic analysis of information systems as a design science. Scand. J. Inf. Syst. **19**(2), 39–63 (2007)
42. Iivari, J., Hirschheim, R.: Analyzing information systems development: a comparison and analysis of eight IS development approaches. Inf. Syst. **21**(7), 551–575 (1996)
43. Jordan, M.I., Mitchell, T.M.: Machine learning: trends, perspectives, and prospects. Science **349**(6245), 255–260 (2015)
44. Karsten, H.: Collaboration and collaborative information technologies: a review of the evidence. DATA BASE Adv. Inf. Syst. **30**(2), 44–64 (1999)
45. Kaufmann, M., Wilke, G., Portmann, E., Hinkelmann, K.: Combining bottom-up and top-down generation of interactive knowledge maps for enterprise search. In: Buchmann, R. et al. (eds.) KSEM 2014. LNAI, vol. 8793, pp. 186–197. Springer (2014)
46. Kaufmann, M.A., Portmann, E.: Biomimetics in design-oriented information systems research. In: Donnellan, B., et al. (eds.) At the Vanguard of Design Science: First Impressions and Early Findings from Ongoing Research Research-in-Progress Papers and Poster Presentations from the 10th International Conference, DESRIST 2015, Dublin, Ireland, 20–22 May, pp. 53–60 (2015)
47. Kim, J.: Making sense of emergence. Philos. Stud. **95**, 3–36 (1999)
48. Kim, J.: Emergence: core ideas and issues. Synthese **151**, 547–559 (2006)
49. Langley, P., Simon, H.A.: Applications of machine learning and rule induction. Commun. ACM **38**(11), 55–64 (1995)
50. Levina, N.: Collaborating on multiparty information systems development projects: a collective reflection-in-action view. Inf. Syst. Res. **16**(2), 109–130 (2005)
51. Luna-Reyes, L.F., Zhang, J., Gil-Garcia, J.R., Cresswell, A.M.: Information systems development as emergent socio-technical change: a practice approach. Eur. J. Inf. Syst. **14**, 93–108 (2005)
52. Lycett, P., Paul, R.J.: Information systems development: a perspective on the challenge of evolutionary complexity. Eur. J. Inf. Syst. **8**, 127–135 (1999)
53. Lyytinen, K.J., Ngwenyama, O.K.: What does computer support for cooperative work mean? A structurational analysis of computer supported cooperative work. Account. Organ. Inf. Technol. **2**(1), 19–37 (1992)
54. Lyytinen, K., Yoo, Y.: Special issue: Issues and challenges in ubiquitous computing. Commun. ACM **45**(12), 62–65 (2002)
55. Macías-Escrivá, F.D., Rodolfo Haber, R., Raul del Toro, R., Hernandez, V.: Self-adaptive systems: a survey of current approaches, research challenges and applications. Exp. Syst. Appl. **40**, 7267–7279 (2013)
56. Markus, M.L., Majchrzak, L.A., Gasser, L.: A design theory for systems that support emergent knowledge processes. MIS Q. **26**, 179–212 (2002)
57. Markus, M.L., Robey, D.: Information technology and organizational change: causal structure in theory and research. Manag. Sci. **34**(5), 583–598 (1988)
58. Merali, Y.: Complexity and information systems. In: Mingers, J., Willcocks, L. (eds.) Social Theory and Philosophy of Information Systems, pp. 407–446. Wiley, London (2004)
59. Minzberg, H.: The Rise and Fall of Strategic Planning. Prentice-Hall, Hemel Hempstead, Hertfordshire, UK (1994)
60. Müller-Schloer, C., Sick, B.: Emergence in organic computing systems: discussion of a controversial concept. In: Yang, L.T., et al. (eds.) ATC 2006. LNCS, vol. 4158, pp. 1–16. Springer (2006)
61. Nan, N.: Capturing bottom-up information technology use processes: a complex adaptive systems model. MIS Q. **35**(2), 505–532 (2011)
62. Ngwenyama, O.K.: Groupware, social action and organizational emergence: on the process dynamics of computer mediated distributed work. Account. Organ. Inf. Technol. **8**, 127–146 (1998)

63. Nielsen, C.B., Larsen, P.G., Fitzgerald, J., Woodcock, J., Peleska, J.: Systems of systems engineering: basic concepts, model-based techniques, and research directions. ACM Comput. Surv. 48(2), 18:1–18:41 (2015)
64. Nijs, D.E.L.W.: The complexity-inspired design approach of imagineering. World Futures 72 (1–2), 8–25 (2015)
65. Orlikowski, W.J.: Improvising organizational transformation over time: a situated change perspective. Inf. Syst. Res. 7(1), 63–92 (1996)
66. Orlikowski, W.J.: Using technology and constituting structures: a practice lens for studying technology in organizations. Organ. Sci. 11(4), 404–428 (2000)
67. Parnas, D.L.: Software aspects of strategic defense systems. Commun. ACM 28(12), 1326–1335 (1985)
68. Patel, N.V., Eldabi, T., Khan, T.M.: Theory of deferred action agent-based simulation model for designing complex adaptive systems. J. Enterp. Inf. Manag. 23(4), 521–537 (2010)
69. Reid, J., Hull, R., Clayton, B., Melamed, T., Stento, P.: A research methodology for evaluating location aware experiences. Pers. Ubiquit. Comput. 15, 53–60 (2011)
70. Salvaneschi, P.: Modeling of information systems as systems of systems through DSM. In: SESoS'16, Austin, TX, USA, pp. 8–11 (2016)
71. Stephan, A.: Varieties of emergentism. Evol. Cogn. 5(1), 49–59 (1999)
72. Symons, J.: Computational models of emergent properties. Mind. Mach. 18, 475–491 (2008)
73. Taleb-Bendiab, A., England, D., Randles, M., Miseldine, P., Murphy, K.: A principled approach to the design of healthcare systems: Autonomy vs. governance. Reliab. Eng. Syst. Saf. 91, 1578–1585 (2006)
74. Tanriverdi, H., Rai, A., Venkatraman, N.: Reframing the dominant quests of information systems strategy research for complex adaptive business systems. Inf. Syst. Res. 21(4), 822–834 (2010)
75. Tarasewich, P.: Designing mobile commerce applications. Commun. ACM 46(12), 57–60 (2003)
76. Thompson, M.P.A.: Cultivating meaning: interpretive fine-tuning of a South African health information system. Inf. Organ. 12, 183–211 (2002)
77. Tolk, A., Diallo, S.Y., Turnitsa, C.D.: Applying the levels of conceptual interoperability model in support of integratability, interoperability, and composability for system-of-systems engineering. Syst. Cybern. Inform. 5(5), 65–74 (2007)
78. Truex, D., Baskerville, R., Klein, H.: Growing systems in emergent organizations. Commun. ACM 42(8), 117–123 (1999)
79. Truex, D.P., Baskerville, R., Travis, J.: Amethodological systems development: the deferred meaning of systems development methods. Account. Manag. Inf. Technol. 10, 53–79 (2001)
80. Truex III, D.P., Klein, H.K.: A rejection of structures as a basis for information systems development. In: Stamper, R.K., Kerola, P., Lee, R., Lyytinen, K. (eds.) Collaborative Work, Social Communications and Information Systems, pp. 213–235. Elsevier (North-Holland), Amsterdam (1991)
81. Ulieru, M., Doursat, R.: Emergent engineering: a radical paradigm shift. J. Auton. Adapt. Commun. Syst. 4(1), 39–60 (2011)
82. Valckenaers, P., Van Brussel, H., Hadeli, Bochmann, O., Saint Germain, B., Zamfirescu, C.: On the design of emergent systems: an investigation of integration and interoperability issues. Eng. Appl. Artif. Intell. 16, 377–393 (2003)
83. van Steen, M., Pierre, G., Voulgaris, S.: Challenges in very large distributed systems. J. Internet Serv. Appl. 3, 59–66 (2012)
84. Varenne, F., Chaigneau, P., Petitot, J., Doursat, R.: Programming the emergence in morphogenetically architected complex systems. Acta. Biotheor. 63, 295–308 (2015)
85. Vogiazou, Y., Raijmakers, B., Geelhoed, E., Reid, J., Eisenstadt, M.: Design for emergence: experiments with a mixed reality urban playground game. Pers. Ubiquit. Comput. 11, 45–58 (2007)
86. Väyrynen, K., Iivari, J.: The competitive potential of IT applications—an analytical-argumentative evaluation. In: ICIS 2016 (2015)

87. Wagner, E.L., Newell, S., Piccoli, G.: Understanding project survival in an ES environment: a sociomaterial practice perspective. J. AIS **11**(5), 278–297 (2010)
88. Wears, R.L., Cook, R.I., Perry, S.J.: Automation, interaction, complexity, and failure: a case study. Reliab. Eng. Syst. Saf. **19**, 1494–1501 (2006)
89. Webster, J., Watson R.T.: Analyzing the past to prepare for the future: writing a literature review. MIS Q. **26**(2), xiii–xxiii (2002)
90. Wegner, P.: Why interaction is more powerful than algorithms? Commun. ACM **40**(5), 80–91 (1997)
91. Wegner, P., Eberbach, E.: New models of computing. Comput. J. **47**(1), 4–9 (2004)
92. Wolf, M.: Embedded software in crisis. Computer **49**(1), 88–90 (2016)

Enterprise Architecture Context Analysis Proposal

Małgorzata Pańkowska

1 Introduction

Generally, context is any information that can be used to characterize a situation of an object. The object is a person, IT product or plan that are considered relevant to the interaction between users. Context has a significant impact on the way humans or machines act, on how they interpret things, and on how they combine their experience together to give it meaning. Generally, the perceived objects remain unchanged, but the perception of them and the relations among them are different. Taking into account the general properties of context, it should be noticed that context is always infinite. The context specification and description details depend on the purpose of why it is done, by whom and for whom. Every entity involved in the context formalization process introduces new backgrounds and perspectives. Context is always dynamic, because the real world is changing beyond the formalization. Context is also considered as a set of constraints that influence the behaviour of an object involved in a given task. An information system is adequate to its context, if the exchanged information is compatible in itself and if the resources required for information processing are available. For computerized application, context is typically the location, identity and state of people, computational and physical objects [1].

In this paper, the category of context information captures the relations an entity has established to other entities, e.g., information systems. Such surrounding entities can be persons, things, devices, services or information. The network of all

---

A prior version of this paper has been published in the ISD2016 Proceedings (http://aisel.aisnet.org/isd2016/proceedings2016).

M. Pańkowska (✉)
University of Economics in Katowice, Katowice, Poland
e-mail: pank@ue.katowice.pl

© Springer International Publishing Switzerland 2017
J. Gołuchowski et al. (eds.), *Complexity in Information Systems Development*,
Lecture Notes in Information Systems and Organisation 22,
DOI 10.1007/978-3-319-52593-8_8

relations and the structure of the related entities construct a context for a particular entity in this network. The structure of relations is changing dynamically, but essentially determines an entity's context. In this paper, context obtains a specific role in communication of entities combined in the network. So, the context is a framework for the EA organizational analysis. Therefore, the EA context is interpreted as a network of stakeholders, a network of principles, and as a network of information systems surrounding the modelled enterprise.

In the paper, the deduction method of thinking is posited, so in the first part of the paper context is defined and widely discussed, its different interpretation are visualized with the mind mapping tool usage. Next, in the second part, the short case study is presented, where a certain types of context are considered and visualized in ArchiMate language.

2 Different Interpretations of Context

Business decision making states the importance of knowledge acquisition in a context. Decision making can be planned as a context awareness system, where tasks and situations are determined by the social environment of the decision maker. It is contrasted with the academic deliberations, where knowledge is out of context, i.e., abstract, de-contextualized.

ISO/IEC 25063 standard provides the Common Industry Format for documenting the context of use for information systems. So, the description of the context of use includes information about the users and all other stakeholders, the characteristics of each user group, the users' goals, their tasks and the environment, in which the system is used. According to the standard, the context description is applicable to software and hardware systems, products or services. It provides a collection of data relevant for analysis, specification, design and evaluation of an interactive system from the perspective of various user groups [2].

Context can be used to decrease impact or enhance existing business measures. Context information is useful for business decision making, so for example:

- Information about the current state: the user's current location, time, activity, people nearby, physiological state, available services, network connectivity.
- User preferences and relationships, including recommendations from people. This type of context information is interesting as it involves personal and social information in making business decisions.
- Accumulated experiences and knowledge, therefore, historical information is used in relation to trust based on previous outcomes.

Beyond that, context can be identified with colours, size, distance, relation details, design, form or background. There is no single definition of context, no single application and no single method. Context enables to know, understand, see and act. For example, mobile phones represent people and their acting. Primary context covers location, activity, time, identity, weather, friend, email address, and phone number. Therefore, computerized systems are able to recognize users and send information to remind about somebody or something, on weather, on social events. Capturing data by sensors and mobile devices can be used for creating context based activities, for monitoring and forecasting the human behaviour [3].

3 Context Considerations in EA Frameworks

There are many frameworks that support the EA modelling and development, however, the context issues are really emphasized in the EA Framework provided by Zachman [4]. The Zachman Framework (ZF) analyses the basic structure for organizing business architecture through dimensions such as data, function, network, people, time and motivation. Zachman describes the ontology for the creation of EA through negotiations among several actors. The ZF presents various views and aspects of the enterprise architecture in a highly structured and clear-cut form. It differentiates between the following levels: Scope (i.e., contextual, planner view), Enterprise Model (i.e., conceptual, owner view), System Model (i.e., logical, designer view), Technology Model (i.e., physical, builder model), Detailed Representation (i.e., out-of-context, subcontractor), and Functioning Enterprise (i.e., user view). In the ZF, the EA context is expressed as the six aspects of the enterprise architecture. The ZF works with the following aspects: Data (what?), Function (how?), Network (where?), People (who?), Time (when?), Motivation (why?). Each aspect interrogates the architecture from a particular perspective. Taken together, all the aspects and some views create a complete picture of the enterprise. In the ZF, the first viewpoint is the planner's view. There are the architect's first sketches and drawings that base on the owner's requirements and the description of an idea what the product, i.e., EA, would look like. On that level, these descriptions would list things important to the enterprise, processes performed by the business, locations where the business operates, organizations important to the business, events significant to the business and the business goals and strategies of the enterprise. They define the scope and boundaries for the enterprise. The plans in the first four viewpoints from the planner's view to builder's view are in context as they describe the product in entity. However, the plans at the component (i.e., Detailed Representation level) are out-of-context as they concern only parts of the total structure. This distinction is significant, because being out-of-context make these components highly reusable; if they are highly standardized, they can be used in many contexts.

Since 1999 the Federal Enterprise Architecture Framework (FEAF) has promoted shared development of business processes and interoperability as well as the sharing of information among US federal agencies and other governmental entities [5]. The FEAF components of an enterprise architecture cover architecture drivers, strategic direction, current architecture, target architectures, transitional processes, architectural components, architectural models, and standards. The architect is responsible for ensuring the completeness of the architecture, in terms of adequately addressing all the concerns of all various views, satisfactory reconciling the conflicts among different stakeholders. The framework emphasizes the role and the view of planner, owner, designer, builder and subcontractor in the EA development process. Therefore, the FEA (Federal Enterprise Architecture) is an attempt to unite some views and functions under a single, common and ubiquitous architecture. Each view is considered as providing a separate context. The FEAF is derived from the Zachman Framework, however, the user of realized architecture is not included in the development team. Planning of enterprise architecture according to the ZF meets some unclear situations (e.g., answer When? is difficult), therefore the FEAF seems to be the simplified and more intense version of the ZF.

The other frameworks of enterprise architecture, although focused on the architectural components development, also include questions concerning the EA views and viewpoints. The Ministry of Defence Architectural Framework (MODAF) is the UK Government specification for architectural frameworks for the defence industry [6]. The MODAF covers seven viewpoints. The All View viewpoint is created to define the generic, high-level information that applies to all the other viewpoints. The Acquisition viewpoint is used to identify programmes and projects that are relevant to the framework and that will be executed to deliver the capabilities that have been identified in the strategy views. The Strategic viewpoint defines views that support the analysis and the optimization of military capability. The intention is to capture long-term missions, goals and visions, and to define what capabilities are required to realize them. The Operational viewpoint contains views that describe the operational elements required to meet the capabilities defined in the Strategic view. This is achieved by considering a number of high-level scenarios, and then defining what sort of elements exist in these scenarios. The Operational views are solution-independent and do not describe an actual solution. These views are used primarily as part of tendering, where they will be made available to supplier organizations and form the basis of evaluating the System views that are provided as the supplier's proposed solution. The System viewpoint contains views that relate directly to the solution that is being offered to meet the required capabilities that have been identified in the Strategic views and expanded upon in the Operational views. There is a strong relationship between the System viewpoint and the Operational viewpoint. The System views describe the actual systems, their interconnections and their use. This will also include performance characteristics and may even specify protocols that must be used for particular communications. The Service-oriented viewpoint contains a view that allows the solution to be described in terms of its services. The Technical viewpoint contains two views that allow all the relevant standards to be defined. This is split into two categories: current standards and predicted future standards [6].

The CIMOSA framework is based on four abstract views (i.e., function, information, resource and organization views) and three modelling levels (i.e., requirement definition, design specification and implementation description) [7]. The four modelling views are provided to manage the integrated enterprise model (i.e., design, manipulation, and access). The role of each view is to filter components out of the model according to given perspective. For the management of views, CIMOSA assumes a hierarchy of business units that are grouped into divisions and plants.

According to The Open Group Architecture Framework (TOGAF), an overall Enterprise Architecture consists of the four subsets, i.e., business, technology, data and application architecture. Beyond that TOGAF includes the following views: Function, Management, Security, Builder's, Data Management, User (and the following physical views), Computing and Communications [8]. In that context, the Architecture Development Method (ADM) is regarded as describing a process life cycle that operates at multiple levels within an organization, operating within a holistic governance framework and producing aligned outputs that reside in an Architecture Repository (AR). Beyond that in TOGAF the Enterprise Continuum provides a valuable context for understanding architectural models. It shows building blocks and their relationships to each other and the constraints and requirements on a cycle of architecture development. In the EA development process, the viewpoints and views ensure a fragmentation and partial specification, however, this approach seems to be useful because of ambiguity and multi-interpretation of context.

4   Network of Stakeholders as EA Context

Shron argues that contexts emerge from understanding who you are working with and why you are doing what you are doing [9]. People learn the context from talking to others. The contexts set the overall tone of the projects, and guide the choices. The generic process of constructing the EA models consists of recognition of the environment of the initiatives, involved stakeholders, organizational culture and management commitment.

Martini and Aloini also argue that EA context is to be extended to cover learning about markets, practices such as lead user experimentation, unconventional tools, openness to external sources, practices that enable the search breadth and idea hunting [10]. Therefore, EA modelling requires studying the environment (see Fig. 1), wherein the business organization is immersed. Anthopoulos and Tougountzoglou [11] for digital city analysis consider a different set of factors. According to them, geographic factors refer to the geopolitical conditions in the country, city or region where the digital city will be located. Economic and market factors refer to wealth, enterprises and growth level in the particular area. The good financial conditions of households and firms support technology and innovative initiatives acceptance. Social factors concern the intention of local community to participate in project planning and project result exploitation. The political factors may support the transparency of public procedures and encourage the project initiation. Legal factors focus on the flexibility and the presence of rules and procedures for e-service deployment and use. Cultural factors concern social attitudes and indicate the existence of communities of common interests.

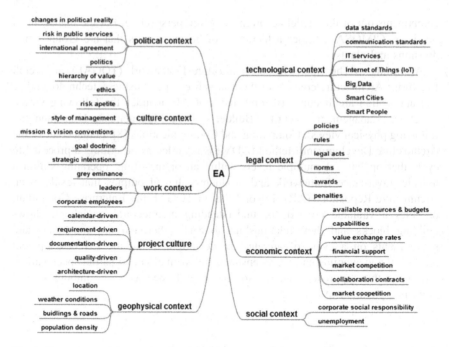

Fig. 1 Enterprise architecture context specification

Duffy [12] specified five different project cultures, which could be also included in the EA project implementation. Calendar-driven culture is characterized by an obsessive focus on schedule, movement from one milestone to the next and all decisions being based on short-term expectations. Requirement-driven culture focuses on functional and non-functional system requirements, and even a small change in the requirement specification is the signal of instability of the planned system. Documentation-driven culture is oriented towards producing the project documentation. The specific challenge is to determine which document to produce next. Quality-driven culture focuses on the quantifiable measures for characteristics such as performance, reliability and security, portability, maintainability, and scalability. The architecture-driven culture is oriented towards an accommodation of the new requirements. In that culture, the user can experiment with different versions of the system for its further incremental and iterative development. That culture supports the construction of adaptable frameworks, which are tuned to suit user requirements [12].

Technological factors refer to the technologies and technology standards that are involved in the EA projects, and to the existence of the appropriate IT industry. Human factors indicate the existence of supervisors and executives with proper skills.

The EA viewpoints define abstractions on the set of models representing the enterprise architecture, where each is aimed at a particular type of stakeholder and addressing a particular set of concerns. According to Lankhorst et al., viewpoints are designed for the purpose of service as a means of communication in a conversation about certain aspects of an architecture [8]. Each viewpoint means a different context. In general, the use of an architectural viewpoint will pass through a number of phases, i.e., scoping, creation of views, validation, obtaining commitment and informing the EA stakeholders. The activities cover generating EA views. The views are primarily the constructs for representing the architecture from different perspectives or viewpoints. The views are very effective as a means for communicating the architecture among the EA stakeholders. In EA frameworks and methodologies, there are different answers to who the stakeholders are.

Generally, in the EA environment, stakeholders need an influence on the EA realization by a number of drivers, e.g., strategy changes, a changing business and regulatory environment, and new technologies. Business managers are interested in business metrics and on reports that highlight some performance measures with the ability to view the same data but through different views. IT engineers are interested in system analysis and look for the metrics to determine the actual cause of critical events, e.g., operation shutdown or random maintenance episodes. Data scientists are responsible for performing ad hoc analysis on a multitude of data sets in heterogeneous systems, leveraging a wide variety of statistical and machine learning algorithms [13]. An obvious way to keep an adequate eye on the interests of stakeholders is a more direct involvement of them in enterprise architecture development activities and the assessment of top management.

The enterprise architects should be able to translate the strategic initiatives and areas of concerns in a concrete enterprise design. The areas important for the enterprise architect knowledge cover system thinking, business and organization, information, information technology, enterprise development and change. The enterprise architect is responsible for documenting, analysing and designing the business processes, business function, products, business units and business objects and the interactions between them. By the analysis of the entire business model, the enterprise architects are able to uncover the points where there is a need for action and the potential for optimization. There is a necessity to ensure the cohesion among all the other roles, i.e., application managers, project managers, process architects, business analysts, IT service providers, IT infrastructure providers, project portfolio controllers, IT strategists, IT managers, security representatives, risk managers, and quality managers (see Fig. 2). However, architecture development requires deep understanding of the enterprise business environment, which cover suppliers, customers, substitutes, government agencies, competitors, and new entrants as it is specified in Michael Porter model (see Fig. 2).

The EA is typically to provide management with an outlook on the coming 3–5 years. The EA facilitates decision making processes by providing a holistic view of the enterprise, leading to better decision making. The enterprise architect is placed in a network of stakeholders (see Fig. 2). They are important only where presence of various diverse interests and elements of negotiations is apparent. Each of them represents a number of interests, which may include the achievement of the whole EA goals. As actors in a network, the EA stakeholders (see Fig. 2) achieve

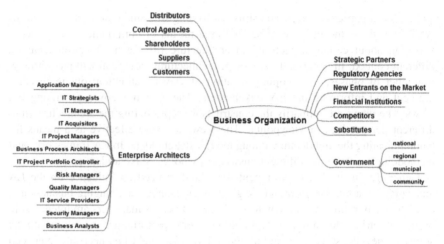

Fig. 2 The EA stakeholders' network

their significance by being in relation to one another. For further consideration, the Actor Network Theory (ANT) developed by Latour and Callon is useful to describe the creation and evolution of socio-technical networks [14]. According to the ANT theory, an actor is defined as an entity making other elements dependent upon itself. The position of the architect in the enterprise determines the associated controls of the EA development activities.

5 Network of IT Systems as EA Context

The stakeholders presented in Fig. 2 belong to certain business units, which dispose certain business information systems. These systems constitute an environment that should be respected in the EA development process. There is an opportunity to apply system context approach to emphasize the value of stakeholder systems. According to Mitra, system context documents how the IT system, which belongs to the analysed enterprise and which is typically represented as a black box, interacts with external entities, i.e., systems of competitors, suppliers, customers, and government agencies [15]. Analysis of the context of other network systems allows to clarify, confirm, and capture the environment, in which the system has to operate. The nature of the external systems, their interfaces, and the information and control flows are inputs to the downstream specification of the technical artefacts in the EA [15]. The System Context provides a catalogue of systems that are external to the system under consideration, the information flow with the external systems, the external events that the Technology System users need to be aware of or respond to, along with a catalogue of profiles of different types of user roles that will be accessing and interacting with the Information System to harness its capabilities [15].

Therefore, business architecture can be defined as the set of structures and stories that underpin "the business of business". However, in each case, the relations among business systems are different, so they are discussed as follows:

- No interaction, e.g., a certain anarchy, because the business organization is centred in itself, without external context.
- Direct transactional interactions, i.e., supply chain, where the suppliers, customers and others are connected in the direct value network.
- Indirect transactional interactions, including market systems of business analysts, recruiters, regulators, standards bodies, competitors in the overall marketplace for this type of enterprise.
- Non-transactional interactions, but creating enterprise ecosystems, including investors, families, communities, non-clients, anti-clients, and others that can be impacted by and impact upon the business organization (see Fig. 3).

John Zachman emphasizes that in the EA development, the environment issues are described through ontologies. Hervas et al., specify three types of ontology [16]. The User Ontology is describing the user profile, their situation, i.e., location, activities, roles and goals, as well as their social relationships. The Device Ontology is the formal description of the relevant devices and their characteristics, associations and dependencies. The third, Physical Environment is defining the space distribution [16]. The enterprise ontology visualisation reveals the relevant elements

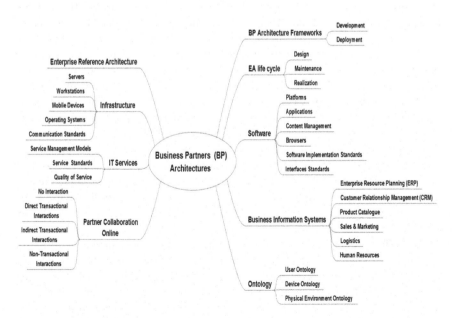

Fig. 3 The EA business partners' systems

of the context as well as metaphors, patterns, pipeline issues, interaction paradigms and methods, view structure, user's social organization, data properties and scalability issues (see Fig. 3).

6 Network of Principles as EA Context

Wasson reminds that IEEE 1471-2000 definition of architectures focuses on the principles guiding the EA design and evolution [17]. Stair and Reynolds [18] think of principles as basic truths or rules that remain constant regardless of the situation. They provide strong guidelines for decision making. For example, practitioners in many disciplines prepare a code of ethics that determine the principles and core values that are essential to their work and govern their behaviour. Usually, principles are based on empirical deduction of observed behaviour or practices. The EA principles are strongly related to goals and requirements. Similar to requirements, principles define intended properties of EA systems. While a requirement states a property that applies to a specific system, a principle defines general characteristics that apply to any system. However, the principles are different for different enterprises and in each case the set of principles is different and as such, that collection of principles constructs the EA context. A principle must be specific for a given EA system by means of one or more requirements or constraints, in order to enforce that the EA system conforms to the principle. The EA principles can be descriptive, explanatory, predictive or prescriptive [19]. The scientific principles are cross-disciplinary and they are applicable in various design domains [20]. They are laws or facts of nature underlying the working of an artefact. The normative principles are based on artefacts such as strategy and influence other business, as well as guidelines, requirements or implementation plans. They are declarative statements that normatively prescribe a property of EA products. The principles are prescriptive because they concern the good practices of EA development, and they are predictive, because they concern the vision of ICT in the enterprise (see Fig. 4).

TOGAF defines an architecture principle as a qualitative statement of intent that should be met by the architecture. In ArchiMate, TOGAF visualisation tool, the Business Model Canvas is a source for motivating architecture principles and it states the business context for the EA description development. The canvas provides nine building blocks to describe the rational of how an organization creates, delivers and captures value. The building blocks are: customer segments, value propositions, channels, customer relationships, revenue streams, key resources, key activities, key partnerships and cost structure (see Fig. 4).

According to TOGAF, principles are general rules and guidelines that inform and support the way in which an organization sets about filling its mission [21]. In TOGAF, principles as inherent laws can be observed and validated and they always concern the stakeholders. In libraries of good practices for IT management and governance the principles useful for EA description development are also hidden. They should be revealed, considered and applied as the EA context. For

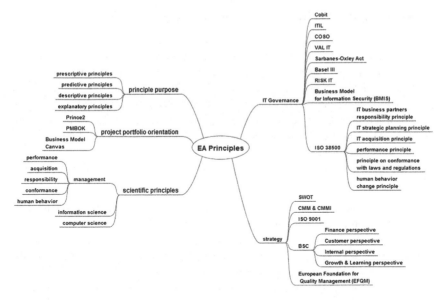

Fig. 4 The EA principles' network

example, Cobit 5 is based on 5 key principles for IT governance and management, i.e., meeting stakeholder needs, covering the enterprise end-to-end, applying a single, integrated framework, enabling a holistic approach, separating governance from management. In the EA development aspect, the principles concerning IT governance as a way of strategic thinking seem to be important. IT governance ensures that stakeholder needs, conditions and solution options are evaluated to determine balanced enterprise objectives to be achieved, setting directions through prioritization and decision making, and monitoring performance and compliance against the established objectives [22]. Compliance with applicable laws and regulations, reliability of financial reporting, and effectiveness and efficiency of operations are also emphasized in COSO (Committee of Sponsoring Organizations) internal control concept, affected by an entity's board of directors, managers and other personnel and designed to provide reasonable assurance regarding the achievement of objectives [13].

The EA principles are a representation of an enterprise as they are embodied in the EA elements and their relationships in an appropriate model. They are fundamentals for description, construction and evaluation of system architecture. According to Sandkuhl et al., they are statements that provide a context for EA modelling and they support the transformation process of an enterprise [23]. The process of the EA principles development covers the following phases: principle identification and formulation, documentation, implementation, monitoring and adjustment. After the deployment, the EA principles should be communicated, regularly monitored and renewed during application.

7 Smart City Architecture Context—Case Study

Understanding context means allowing for familiarization with the nature of the business and sharing the information among the EA development team members. The EA development project is then considered in the perspective of the overall organization, which determines the project goals. The EA team needs to consider context when choosing which processes, practices and techniques they use and therefore they are sure they are doing the right things and they are not doing things that are unnecessary [24].

In the EA development process the context knowledge is explored in some subprocesses:

1. The EA project goals specification, context scope determining, taking into account appropriate views and viewpoints of the EA.
2. The EA knowledge contextualization, what means enriching the acquired knowledge with contextual information from different sources and about different entities included in the EA description.
3. The EA knowledge re-contextualization, what means creating knowledge through the sharing of experiences in the project team and monitoring the different sources of context. The re-contextualization is an iterative process. During which knowledge from different sources is evaluated and explored in the further plans and implementations.
4. The EA knowledge de-contextualization, which means knowledge abstracting and generalization in the computerized information system designing process.
5. Although the implementation of technical solution can be context free, the further deployment should be context sensitive, because it is realized in a certain social environment.

The context description for the EA development requires specifying the context scope, which determines what must be (or should) be considered for EA modeling and implementation. The context contents and its scope are relative and depend on the stakeholders involved in the EA modeling process. From the EA constructivist point of view, context scope and context knowledge are presented in the documentation of the project. The EA contextualization requires a critical reflection of the social and historical background of business activity, because the EA developer should see how the current behavior of citizen in the residential environment is emerging.

In the presented below case study, architecture description focuses on modelling the system architecture for the garbage collection in a municipality supported by mobile technology. The analysed problem belongs to the IT solutions for smart cities.

The smart city system architecture is realized in the circumstances of strong connection among IT governance and municipality strategy. The smart city architecture modelling starts with modelling of IT resources, i.e., hardware, software and networks as well as with modelling of the business processes and governance principles selection. The smart city architecture modelling is located in the city planning and formulated taking into account an analogy between city and system architectures (see Table 1). Taking into account the Table 1 content, the municipality planning can be considered as a certain waypoint for smart city architecture development and it is a context for city IT architecture. Therefore, knowing the city the designers of Waste Management System can develop, implement and deploy the system easily. The project starts from project goals' specification, project feasibility recognition, and context scope identification. Wide spectrum of context knowledge encourages to reduction of the context aspects and to selection the most important issues. So, the EA developers are assumed to consider behaviours of citizens and city visitors, culture of waste management, demographics, local government priorities and principles, garbage collection transportation infrastructure, seasons of year and weather.

Table 1 Analogy between municipality planning and city IT architecture development

Comparison criteria	Municipality planning	City IT architecture development
Advisability	The city planning to satisfy the citizens' needs and requirements	Development of business organization for garbage collection to satisfy the citizens requests
Efficiency	Development of the city so that logistics, supply and demand can be realized economically	Develop a municipality architecture that support efficient operations of garbage collecting
Predictability	City planning for the development of an additional suburbs and their requirements	Ability to forecast the future development of the city in the aspect of demand for the garbage collecting services
Sustainability	Development of the municipality in a sustainable, citizen friendly way	Development of the smart city architecture that is sustainable and complies with legal acts, technology and regulatory standards. Providing long-term IT solutions, considering the modern technology
Scalability	Development of the city to cope with peaks and growth in the city communication and transport	Development of flexible collectors of wastages, so that they can handle the business activity peaks.
Quality of life	Ensuring a high quality life for the citizens	Development of an enterprise architecture that allows satisfactory and reliable fulfilment and motivation
Heterogeneity of municipality resources	Infrastructure and building assets are created by different designers and implemented in different time	ICT resources from different hardware and software providers, lack of interoperability and necessity to ensure compatibility and integration

Source Own work based on [25]

The contextualization is realized in the aspect of the EA developers, however, that process is repeated, i.e., the re-contextualization is ensured by the other computerized system developers, designers, software engineers, and managers responsible for deployment. The context consideration and visualization in the EA model is validated in the iterative process of application development, business–IT alignment works, in experimentation process and through discovering new opportunities during mobile application development.

Although, the ArchiMate language express ideas on a high level of planning, the EA developers can use ArchiMate models to visualize assumptions, define desired business impacts and analyse user needs. The EA developers can use the ArchiMate models to drive stakeholder alignment and support prioritization of the needs. The stakeholders can discuss the various deliverables necessary to achieve a specific outcome and determine which ones will play the biggest part. For making good investments, the EA developers can identify options, compare solutions and control their research efforts.

Figures 5, 6 and 7 cover the EA model of wastage collecting in a municipality.

ArchiMate 3.3.2 model consists of elements belonging to each of the following layers, i.e., Strategy, Business, Application, Technology, Physical, Implementation and Migration. In the presented in Fig. 5 model, the Business layer includes:

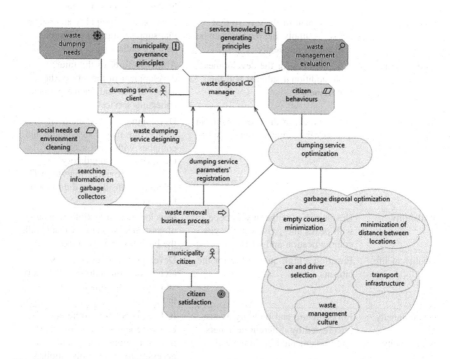

Fig. 5 Waste management ICT architecture model: motivation and business layers

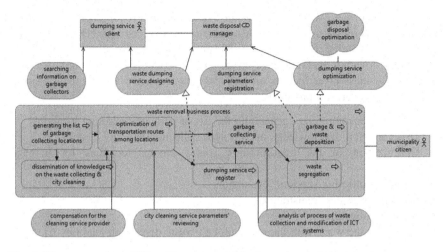

Fig. 6 Waste management ICT architecture model: business process in business layer

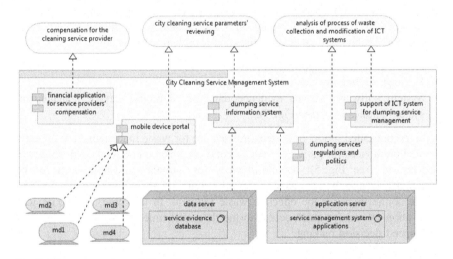

Fig. 7 Waste management ICT architecture model: application and technology layers

- Actors, i.e., municipality citizen and dumping service client.
- Services, i.e., dumping services, including the searching of information on garbage collectors, dumping service designing, service parameters' registration and reviewing, dumping service optimization, compensation for dumping service providers, dumping process analysis.
- Business processes, i.e., waste removal process (see Fig. 5).

The Motivation layer in Fig. 5 concerns the EA business strategy issues, i.e., EA driver, e.g., waste dumping needs, EA goal, e.g., citizen satisfaction, EA principles, e.g., municipality governance principles, service knowledge generating principles,

EA assessment, e.g., waste management evaluation, EA requirement, e.g., social needs of environmental cleaning, EA constraint, e.g., citizen behaviours, (see Fig. 5). The specific context in the IT architecture is visualised by the icon "meaning" covering the knowledge on the garbage collecting process optimization. The EA Business layer focuses on business process, i.e., waste removal process covering the seven sub processes:

- Package generating, i.e., generating the list of garbage collecting locations.
- Dissemination of knowledge on wastages collecting and city cleaning.
- Optimization of the transportation routes among locations for garbage collecting.
- Garbage collecting services' evaluation.
- Dumping services' registration.
- Waste segregation.
- Finalizing the work with the package, i.e., waste deposition (see Fig. 6).

In the EA models, Application layer includes the following components: financial application for service provider compensation, mobile device portal, dumping service information system, dumping services' regulations and politics, support of ICT system for dumping service management. In the smart city architecture, the Technology layer comprises the following components:

- Nodes: data server, application server.
- Devices: mobile devices (md_1 ... md_n).
- System software: service evidence database, dumping service management system applications (see Fig. 7).

In the presented model of system architecture for waste management in the city, the mobile devices for garbage collectors and optimization of waste deposition are emphasized (see Fig. 5). Mobile devices for garbage collectors seem to be important, because they are used to store the context data on dumping location and enable connections for the effective waste removal. In this case, mobile devices application allows for quick access to the data, real time information control and high speed decision making.

8 Conclusion

The smartness of the city is expressed by the number of intelligent buildings, cars, development of transportation infrastructure supported by ICT systems, as well as by the level of implementation of the system for supporting these assets, therefore, the Waste Management System can be placed among the other IT solutions. Modelling of the smart city context requires detailed specification of the modelling aspect and precise explanation if the subject of consideration concerns city economy, mobility of citizens, access to public services, reduction of wastage, and social capital development. The ArchiMate 3.3.2 tool supports the visualization of the EA context by the language elements of all layers, i.e., Strategy, Business, Application, Technology, Physical, Implementation and Migration. However, particularly important is the Business layer, where the "Principle" and "Meaning" elements (e.g., garbage disposal optimization in Fig. 5) are proposed to represent the context knowledge in the EA model. However, the language should be further developed to include the considerations on the EA context, which was in this paper visualised.

References

1. Zimmermann, A., Lorenz, A., Oppermann, R.: An operational definition of context. In: Kokinov, B., Richardson, D.C., Roth-Berghofer, T.R., Vieu, L. (eds.) Modeling and using context, pp. 558–573. Springer, Berlin (2007)
2. ISO/IEC 25063:2014, http://www.iso.org/iso/. Accessed May 2012
3. Ben Mena, T., Bellamine-Ben Saoud, N., Ben Ahmed, M., Pavard, B.: Towards a methodology for context sensitive systems development. In: Kokinov, B., Richardson, D.C., Roth-Berghofer, T.R., Vieu, L. (eds.) Modeling and Using Context, pp. 56–68. Springer, Berlin (2007)
4. Zachman, J.A.: Frameworks standards: what's it all about? In: Kappelman, L.A. (ed.) The SIM Guide to Enterprise Architecture, pp. 66–70. CRC Press, Boca Raton (2010)
5. Federal Enterprise Architecture Framework: Version 1.1, September (1999), CIO Council. http://www.cio.gov/documents/fedarch1.pdf. Accessed May 2012
6. Perks, C., Beveridge, T.: Guide to Enterprise IT Architecture. Springer, New York (2003)
7. Spadoni, M., Abdmouleh, A.: Information systems architecture for business process modelling. In: Saha, P. (ed.) Handbook of Enterprise Systems Architecture in Practice, pp. 366–382. Information Science Reference, Hershey (2007)
8. Lankhorst, M.: Enterprise Architecture at Work. Springer, Berlin (2005)
9. Shron, M.: Thinking with Data, How to Turn Information into Insights. O'Reilly, Beijing (2014)
10. Martini, A., Aloini, D.: Unpacking exploratory innovation: search practices, organizational context and performance. In: Augsdorfer, P., Bessant, J., Moslein, K., von Stamm, B., Piller, F. (eds.) Discontinuous Innovation, pp. 105–139. Imperial College Press, London (2013)
11. Anthopoulos, L.G., Tougountzoglou, T.E.: Viability model for digital cities: economic and acceptability factors. In: Reddick, Ch., Aikins, S.K. (eds.) Web 2.0 Technologies and Democratic Governance, pp. 79–98. Springer, New York (2012)
12. Duffy, D.: Domain Architecture, Models and Architectures for UML Applications. Wiley, Chichester (2004)
13. Tarantino, A.: Manager's Guide to Compliance, Sarbanes-Oxley, COSO, ERM, COBIT, IFRS, BASEL II, OMB A-123, ASX 10, OECD Principles, Turnbull Guidance, Best Practices and Case Studies. Wiley, Hoboken (2006)
14. Callon, M., Latour, B.: Unscrewing the big leviathan: how actors macro-structure reality and how sociologists help them to do so. In: Knorr-Cetina, K.D., Cicourel, A.V. (eds.) Advances in Social Theory and Methodology: Towards an Integration of Micro and Macro-Sociologies, pp. 203–277. Routledge and Kegan Paul, London (1981)
15. Mitra, T.: Practical Software Architecture, Moving from System Context to Deployment. IBM Press, Old Tappan (2016)
16. Hervas, R., Nava, S.W., Fontech, J., Casero, G., Laguna, J., Bravo, J.: Exploring context semantics for proactive cooperative visualization. In: Luo, Y. (ed.) Cooperative Design Visualization and Engineering, pp. 52–55. Springer, Berlin Heidelberg (2009)
17. Wasson, Ch.: System Analysis. Design and Development, Hoboken (2006)

18. Stair, R., Reynolds, G.: Principles of Information Systems, Course Technology. Cengage Learning, Walldorf (2010)
19. Kazem Haki, M., Legner, Ch.: New avenues for theoretical contributions in enterprise architecture principles—a literature review. In: Aier, S., Ekstedt, M., Matthes, F., Proper, E., Sanz, J.L. (eds.) Trends in Enterprise Architecture, Research and Practice Driven Research on Enterprise Transformation, pp. 182–197. Springer, Berlin (2012)
20. Greefhorst, D., Proper, E.: Architecture Principles, The Cornerstones of Enterprise Architecture. Springer, Berlin (2011)
21. Op't Land, M., Proper, E., Waage, M., Cloo, J., Steghuis, C.: Enterprise Architecture, Creating Value by Informed Governance. Springer, Berlin (2009)
22. COBIT 5. A Business Framework for the Governance and Management of Enterprise IT, An ISACA Framework. ISACA, Rolling Meadows (2012)
23. Sandkuhl, K., Simon, D., Wisotzki, M., Starke, Ch.: The nature and a process for development of enterprise architecture principles. In: Abramowicz, W. (ed.) Business Information Systems, pp. 260–272. Springer, Heidelberg (2015)
24. McDonald, K.J.: Beyond Requirements, Analysis with the Agile Mindset. Addison Wesley, New York (2016)
25. Ahlemann, F., Legner, Ch., Schafczuk, D.: Introduction. In: Ahlemann, F., Stettiner, E., Messerschmidt, M., Legner, Ch. (eds.) Strategic enterprise architecture management, pp. 1–33. Springer, Berlin (2012)

Gossip and Ostracism in Modelling Automorphosis of Multi-agent Systems

Mariusz Żytniewski

1 Introduction

The basic element of automorphosis is the organisation of a system. Organisation should be understood here as relation of a system's structure with its function [4]. The structure of a system defines the relationship of its elements as a whole in the manner determined by the system. This manner means that the elements enter into relations with each other, creating a specific sub-system operating as part of the main system. They are connected with each other and integrated. At the same time, the elements of a system are separated from each other as part of subsystems in order to prevent possible interferences. Such dependence of a system's elements is static in a specific point in time and changes with subsequent iterations. As a result, the relations that are created between a system's elements are changeable, and elements of a given sub-system may become elements of a different sub-system. However, this process cannot be random.

The element that impacts the change of the structure of sub-systems is a function. A function defines the objectives of a system and dynamically created sub-systems, impacting relations between its elements and defining them. A change of the objectives of the main system, and consequently its functions, leads to change of the relations between its elements, and as a result—change of the structure of sub-systems.

Thus, it can be indicated that automorphosis of a system results from accepted objectives of its operation, i.e. the functions it fulfils. These objectives can be

A prior version of this paper has been published in the ISD2016 Proceedings (http://aisel.aisnet. org/isd2014/proceedings2016).

M. Żytniewski (✉)
University of Economics in Katowice, Katowice, Poland
e-mail: mariusz.zytniewski@ue.katowice.pl; zyto@ue.katowice.pl

© Springer International Publishing Switzerland 2017
J. Gołuchowski et al. (eds.), *Complexity in Information Systems Development*,
Lecture Notes in Information Systems and Organisation 22,
DOI 10.1007/978-3-319-52593-8_9

defined through mechanisms of a system in which sub-systems are built or through the elements of the sub-systems themselves. It can be assumed here that all functions of a sub-system are regulated by the system in which such a sub-system is being build, whereas within sub-systems the relations result from the objectives of the units that compose them.

In the case of IT systems, the term automorphosis indicates systems' capability of independent creation of their internal structure, not in a random way, but a coordinated one and aimed at achievement of defined objectives. When creating IT systems, automorphosis of systems requires that they have a specific construction connected with the necessity to divide the system into coherent elements that have various functionality and are capable of cooperation through defined mechanisms of communication. Examples of such solutions include multi-agent systems, in particular software agent societies [13–15]. Software agent societies are treated as a sub-type of multi-agent systems whose fundamental characteristic is openness expressed in the capability of joining the structure of a society of new agents and dynamics connected with lack of clearly accepted relationships, e.g. hierarchical ones, between agents. Thus, a software agent society requires mechanisms that protect their operation and support coordination of the creation of groups of task agents designed to perform specific actions of an agent system.

Earlier research conducted by the author was focused on pointing out that a software agent society, in the context of modelling their structure, can be viewed as a metaphor of a social organisation in which an agent society can support an organisation's actions through delegation of its specific tasks of business process participants [18, 19]. The research was focused on defining possibilities of involving a process of modelling the operation of a social organisation in the process of modelling the architecture of software agent societies, which would contribute to better understanding of tasks of the agents in such a society. As an effect, a methodology was proposed [16] that combined a process approach and its modelling in Business Process Modeling Notation (BPMN) with the possibility of connecting an organisation's processes with the architecture of an agent society.

A certain limitation of the developed methodology for designing an agent society was the problem of defining the mechanisms of a society's automorphosis. While available methodologies for designing multi-agent systems enable presentation of their static structure and indication of possible states of a multi-agent system, in the case of the features of openness and dynamics that characterise societies, these methodologies do not enable illustration of control mechanisms that software agent societies have to possess [20]. For that reason, the author initiated research connected with the possibility of using trust mechanisms as an element supporting control of the operation of an agent society [17]. A complete list of papers on this subject can be found on the website www.projektncn.katowice.pl.

As will be pointed further in the paper, reputation and trust defined in an agent society can be extended to include the use of the concepts of gossip and ostracism.

2   Current State of Research on the Use of Gossip and Ostracism in IT Systems

Software agent societies are one of the concepts of building multi-agent systems. The main assumptions of the development of such solutions are openness of a system, i.e. possibility of agents joining and leaving a society, and focus of a system on examining agents' impact on each other and their environment. The first of the above-indicated characteristics, referring to openness of a system, leads to the necessity of modelling automorphosis of such a system. The variety of agents remaining in a system at a given moment and changing needs of a system make it necessary to analyse the functionality of units, match them to tasks and to implement the mechanisms for coordination of assignment of tasks to agents in a society. A system openness as well as mobility of agents contribute to the transitory nature of relations established between agents. Thus, a business process performed at a given moment by a system may be performed in another iteration by different agent units. It may also lead to the situation where several agents will strive to participate in the performance of a specific task in a given iteration. These problems make it necessary to apply mechanisms enabling assessment of agents' activities and are connected with the possibility of using mechanisms for assessment of trust and reputation in such systems [17]. Trust and reputation mechanisms aid the process of matching agents to processes performed by a system. However, they are not sufficient in the context of analysis of a system behaviour. As for the development of software agent societies, their use may be connected with the mechanism of gossip [8] and ostracism [11]. These concepts, widely used in human societies, can be applied as an element that supports control of software agent societies. Research on human societies shows that a human being attaches greater importance to gossip than facts [10].

The first of the mechanisms refers to information propagation. The first application of the mechanism of gossip can be found in the work [2] where it was used to support the mechanism of preserving homogeneity of replicated data bases. The mechanism of gossip is an important element of the activity of a society, as it allows its members to express their opinions. When applying this mechanism in respect of IT solutions, it is necessary to specify a range of assumptions of its use. These are:

- Algorithm for gossip propagation in a system.
- Communication protocol of a system's entity.
- Style of the operation of a communication protocol.

The above-listed elements will be described in the context of their use as part of a software agent society.

Research on algorithms for gossip propagation in a system (without clear specification of its type) refers to its three main types [3]:

- **Information Spread**: this type of algorithms can be used in a situation when one element of a system possesses a specific piece of information and wants to propagate it among other elements of the system.
- **Computing Aggregates**: this type of algorithms can be used in the case of the necessity to propagate the information possessed by individual elements of a system in such a way that an assumed function of aggregation of this information is common to all its elements.
- **Overlay Management**: the last type of algorithms refers to a situation of incomplete knowledge of a system's elements about the existence of other elements in this system. Here, the problem is possibility of propagation of a given piece of information in an indirect way through successive elements of the system.

The research on algorithms indicated herein does not directly refer to the aspect of agent systems, but rather to efficiency and effectiveness of the operation of the process of propagating a given piece of information, and can be used in the process of controlling the flow of information in a multi-agent system. This results from the fact that these algorithms are oriented towards the aspect of time and possibility of propagating a given piece of information in all or at least a selected group of elements of a system, which may constitute a problem in the case of dynamic changes in the structure of a system. In particular, these algorithms refer to structures built as Peer to Peer solutions. As a result, they are largely oriented towards the aspect of a system's structure, rather than its function, which was indicated earlier as the fundamental element of automorphosis. The solution that will be presented further in the text uses the first type of algorithms, which is oriented towards propagation of gossip between those asking an agent and other agents and between a multi-agent platform and an agent.

From the perspective of gossip propagation in a system, it is also necessary to specify the protocol of information exchange between elements of a system. The model presented further in the text used a general protocol for handling gossip as proposed in the work [5]. The reason for using this protocol of information exchange was easiness of its adaptation to Foundation for Intelligent Physical Agents (FIPA) protocol used in the applied multi-agent platform.

The last element is style of the operation of a communication protocol. Literature [1] presents three styles of the operation of gossip defining protocols: dissemination (rumour-mongering) protocols, anti-entropy protocols for repairing replicated data, protocols that compute aggregates, or that accomplish some task as a side-effect of computing an aggregate. In the developed model, the application of the first type was adopted, in particular its sub-type: background data dissemination protocols. This results from the fact that in the case of a multi-agent platform that possesses information about entities of agents existing within it, it is not necessary to use more complex mechanisms. The mechanism is characterised by a periodical mechanism of propagating gossips in a system.

In agent societies, agents' new knowledge can be defined through their mechanism of artificial intelligence which generates a message or through information coming from the environment of a multi-agent platform. Such information, remembered by an agent unit, is treated as belief. This is due to continuous change in an agent's environment, as a result of which information possessed by an agent can be out of date at a given point in time. The second reason for such treatment of an agent's knowledge is possible lack of confirmation of the information it receives from its environment. When communicating with its environment, an agent is unable to verify all the information, which causes it to operate in the conditions of uncertainty, as it tries to perform its actions in a given society. As a result, the information it receives can be treated by it as gossip. It is necessary to distinguish here two types of information acquired by an agent. Information from the platform in which it resides and that from other agents. In the first case, the information acquired by an agent does not have to be confirmed by it, as it is formulated by the mechanism controlling the multi-agent platform which by definition will strive for proper operation of the society. In the second case, information about other agents can be treated by a given unit as gossip defining information about other agents. In this case, an agent receiving a given piece of information has to rely on its trust in the agent that has formulated this piece of information and on trust in the agent to which the gossip refers.

The second mechanism that can be used in an agent society is the mechanism of ostracism. Literature distinguishes three types of ostracism [12]:

- **Punitive Ostracism**—in this approach, a group of units or a single unit stops cooperating with another unit or group as a result of actions taken by it/them.
- **Defensive Ostracism**—where a group or unit refuses cooperation because of fear.
- **Oblivious Ostracism**—where a unit or group is excluded without clearly specified intentions/reasons.

In the context of the earlier-addressed issues concerning automorphosis, structure and function of a system, the first of the presented approaches seems reasonable. The mechanism of ostracism in an agent society is connected with the necessity of supervision and control of agents' behaviour towards each other and in relation to the mechanisms controlling a multi-agent platform. In the case of communication in an agent society, there are two causes of ostracism. The first one concerns formulation of messages transmitted between agents. If a given agent unit detects that a message (which can be treated as gossip) is not true, it can decrease its trust in the agent propagating this piece of information. Decrease of an agent's trust in another unit may lead to the use of the mechanism of ostracism, where an agent ceases cooperating with a given unit due to low trust in it. The second reason for the use of ostracism is not so much propagation of untrue information as a given agent's lack of skills. In the proposed model [6] trust in agents is built through their proper performance of tasks assigned to them. If a given agent unit proposing to perform a specific task performs it incorrectly, the level of trust in it will decrease. As a result, this agent will no longer be taken into account during formation of a group of agents to perform a specific task. Decreased trust in a specific agent's skills can be propagated by agents through the mechanism of reputation [6] (propagated by a multi-agent platform) or, as was already indicated, through the mechanism of gossip (propagated by an agent).

As was pointed to, the mechanisms of gossip and ostracism support the process of controlling the activity of agent societies. One of the works where these mechanisms were addressed is [9]. In the proposed model, the authors made an attempt to define the parameters of an agent society, indicating the following parameters as reasonable [9]—cooperativeness value: this attribute concerns how cooperative an agent is, tolerance value: which characterizes how much non-cooperation the agent can tolerate before it decides to leave the group, rejection limit: how many rejections the agent can face before it decides to leave for another group, gossip blackboard length: a gossip blackboard of certain length to store the gossip messages from other agents of its group, life span: determine how long the agents remain in the society and cost and benefit for sharing.

So defined parameters of agent behaviour control involve only the aspect of communication and do not address the problem of tasks performed by agents in a society. Thus, the approach proposed by the authors focuses on the aspect of cooperation in the context of information exchange without clear specification of the fundamental element of automorphosis, i.e. the already mentioned function of a system.

The solution, proposed later in the paper, constitutes developed research, carried out by the author, on building software agents societies [12–19] and is a proposal to extend the Java Agent DEvelopment Framework (JADE) platform functionality (which is one of the most functional multi-agent platforms [21]) with trust, reputation, gossip and ostracism elements. In the proposed approach the mechanism of agent's action is based on its reactive and proactive layer [22], which interaction allows to control the operation of a unit. As a result, a trust and reputation concept model was developed [17] and indicators, which can be used in the assessment of community agents [6], were pointed. In the proposed paper, this model has been implemented for the JADE platform and extended with additional elements. As a result, the following indicators governing agents' activities can be identified in the proposed trust and reputation model extended to include the mechanism of gossip and ostracism:

- **Level of an agent's cooperation**—defines the number of tasks that it was able to perform in a society relative to all the tasks that had to be performed.
- **Level of an agent's tolerance**—the number of tasks that an agent will propose to potentially perform and that can be rejected before an agent decides to leave the society.
- **Time of an agent's activity**—how long an agent waits for appearance of tasks that it can perform from the moment the last of such tasks appeared. After this time an agent leaves the society.
- **Agent's self-trust**—indicates level of trust that an agent has in its actions. An agent generates it based on its abilities connected with performance of actions

and tasks as part of processes in a society. It is deterministic in character, as the model does not take into account a situation when an agent deceives itself.

- **Social trust in agent n**—defines the level of trust that an agent has in other agents. It is generated based on interaction with other agents connected with actions that they perform. Every agent stores a number of values of this parameter. The number depends on interactions with other agents.
- **Gossip social trust of agent n**—It is used when a new agent appears on platform and specific agent has to adopt gossip trust about other agent.
- **Reputation of agent n**—defines the level of agents' reputation on the platform. It is used when agent does not have social trust in a specific agent unit. It is used to eliminate the problem of "cold start" where an agent that does not have knowledge about other agents will not be able to make a decision about further actions.
- **Gossip reputation of agent n**—It is used when a new agent appears on platform and platform has to adopt gossip about agent reputation from another agent.

For illustration of the operation of the so defined parameters, the paper will present an example of the operation of a society developed based on the author's developed solution. This solution will present how a mechanism of gossip reputation can influence on a society.

3 Proposal to Extend the Model of Trust and Reputation

The model of trust and reputation presented in work [6, 17] can be extended to include the approach connected with gossip propagation and ostracism as indicated herein. In the presented model, a range of trust and reputation parameters were defined and divided into 3 separate groups. The first group of indicators refers to an agent's self-trust, the second group refers to an agent's social trust, whereas the third one—to an agent's reputation in an agent platform. Each of the indicators is examined in terms of four levels, defined as trust/reputation of an action, a task, a process and general one. As a result, indicated sets of agents form a two-dimensional matrix presented in Table 1.

Based on the model [6] which was extended to include the concepts of gossip and ostracism, it can be said that the proposed indicator of cooperation [9] should

Table 1 Trust and reputation matrix developed in accordance with [6]

Level/type	Agent's self-trust	Social trust in agent n	Reputation of agent n
Action	0–1	0–1	0–1
Task	0–1	0–1	0–1
Process	0–1	0–1	0–1
General	0–1	0–1	0–1

result from an agent's skills, i.e. an agent with knowledge in a given field or ability to perform a task should show a higher indicator of cooperation. An agent's tolerance for non-cooperation is connected with the tasks that are propagated in the system. Temporal limit of participation in a society was also indicated in the project, i.e. time limit that an agent will spend in a society before it leaves it. It is counted from the moment of receiving information about a task that can be performed by a given agent.

For the implementation of the so defined model of a society, the JADE multi-agent platform was used as the basis for building software agents. The functionality of the platform was extended to include elements of the mechanism of trust and reputation control as well as the proposed indicators related to gossip generation and ostracism. The developed solution, in accordance with the adopted methodology of designing an agent society, involves the development of a process or a group of processes that should be supported by agents. In accordance with the trust and reputation model extended to include the elements of gossip and ostracism parameters, the following research methodology was adopted:

Parameters of the mechanism for controlling an agent society:

- Defining a set of business processes in which agents will participate.
- Defining a set of agents participating in a system.
- Defining initial indicators of agents' reputation in a system.
- Defining the limit indicator of ostracism of a system.

Parameters of agents that can become an element of a dynamic society:

- Defining skills/tasks of agents that they are able to perform.
- Defining the probability of an agent's performance of a task.
- Defining the level of an agent's ostracism.
- Defining the base self-trust of an agent.
- Defining the base level of social trust in agents.
- Defining the base level of an agent's reputation on the platform.

Parameters defined in this way were imported to a prepared simulator of a society behaviour, and a simulation of the operation of an agent society was carried out.

4 Observed Mechanisms of the Operation of a Society

The developed simulator enables definition of a business process in which an agent society will participate. The defined processes are imported to the simulator of the operation of a society. Next, sets of agents are defined and assigned basic parameters, such as definitions of tasks that they can support and base values of indicators that are necessary to carry out the simulation. Thus, performing simulation consisted in defining the following stages:

1. Developing a process model in BPMN notation—the business process to be performed by a society of agents was modelled in a specially prepared BPMN editor.
2. Preparing a multi-agent platform—the multi-agent platform was developed based on a JADE platform. The functionality of the platform was extended to include agents that process data about agents' reputation and trust.
3. Preparing agents and making them show necessary behaviour—agents capable of performing tasks ascribed to a business process were added to the platform as part of the simulation, and the probability of the performance of the tasks was defined.
4. Providing the platform with a business process to be performed—a business process was presented to the platform and analysed. Next, the currently present agents were notified about the possibility of participation in the process.
5. Agents' response—upon analysis of received messages each agent could apply for participation in the process.
6. Choosing agents—upon receipt of agents' applications and based on data about trust and reputation from previous iterations or based on the gossip mechanism, the multi-agent platform chose agents to perform tasks in a process.
7. Performing a process—the agents chosen by the platform formed a sub-society and performed tasks required by a process.
8. Assessment of iterations—after completing a single cycle the agents were assessed and the relevant indicators were updated. After this stage, the status of agents could be updated and the process of simulation was repeated.

Results will be presented based on one of the prepared processes. The process is illustrated in Fig. 1.

The process consists of three tasks. First the selected agent saves the document (task 1), then the information on saving the document is recorded in the database (task 2), and the user is informed of this action (task 3). For each task two agent actions need to be done. Based on the prepared simulator, the tasks were assigned a

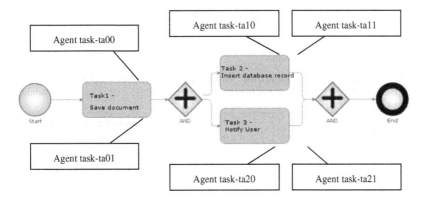

Fig. 1 Analysed business process

set of three agents, each of which can perform a selected task. Prepared simulator was developed as the extension of JADE platform. In the simulation process, each agent behavior is monitored and its progress is recorded in the database. The simulator can operate in two modes. The first is the actual implementation of a business process, where the behavior of agents can be programmed, and the agent society oversees the process. The second one is a simulation mode, where the agents' behaviours are simulated. The probability of correct execution of the task results from the assumed probability level. Figure 2 presents the interface of a developed simulator.

Adopted initial assumptions are presented in Table 2.

The Table 2 shows a set of initial parameters for a single process, assuming a cold start. In this approach, only with the result of the activities of agents it is possible to indicate the value at a given point of time. The following figures will present the changes in the value of reputation for the individual agents and their actions.

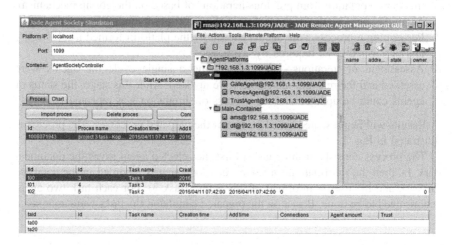

Fig. 2 Agent Society Simulator interface

Table 2 Adopted initial assumptions

Level/type	Agent's self-trust (initial)	Social trust in agent n	Reputation of agent n (initial)	Possible agent	Success probability
Action ta00	0	0	0	ServiceAgent01	0.7
Action ta01	0	0	0	ServiceAgent01	0.7
Action ta10	0	0	0	ServiceAgent02	0.7
Action ta11	0	0	0	ServiceAgent02	0.7
Action ta20	0	0	0	ServiceAgent03	0.7
Action ta21	0	0	0	ServiceAgent03	0.3

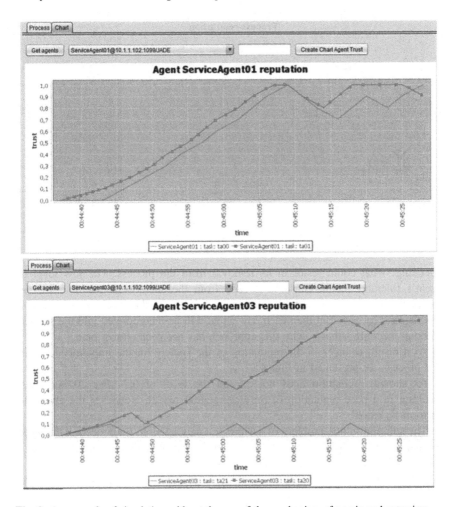

Fig. 3 An example of simulation without the use of the mechanism of gossip and ostracism

For so prepared simulation, lack of the use of the mechanism of gossip and ostracism was first assumed. It was assumed that the simulation will cover 20 iterations of the performance of the indicated process.

As shown in Fig. 3 reputation level of ServiceAgent03 for the action ta21 is very low. This is due to the low probability of the performance of the ta21 task. In the case of ServiceAgent01 the reputation is growing.

Originally used mechanism does not use the indicated mechanisms of gossip and ostracism. As a result, each new agent fed to the platform has a level of confidence at the same level (in the adopted simulation 0). This causes difficulty in the initial selection of agents for the process, because the introduction of a new agent to the process can only take place after the agent reaches a certain level of reputation. In the indicated example ServiceAgent01 has reached 0 reputation after 5s of

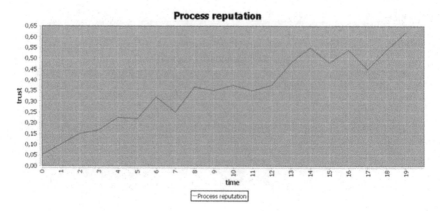

Fig. 4 Example of process reputation

operating, what could result in its removal from the process. The low level of agent's tolerance could cause its escape from the society, despite its skills to complete the task. One certain solution here is to set the level of new agents' reputation level at 1. But then each new agent will be incorporated into the process without checking—that is why this approach has not been shown here.

As a result, changes in the reputation of the society for the implementation of this process are presented in Fig. 4.

As shown, reputation does not exceed the level of 0.7 after 20 iterations, resulting in a poor reputation of ServiceAgent03, which alone has the ability to perform task ta20.

The solution to this problem of assigning a new agent to the process is the used mechanism of gossip and ostracism, allowing to determine the base level of agent's trust based on the knowledge of other units.

After 10th stage, a new agent capable of participating in the operation of the society will be introduced to the platform. The agent, due information about it, will join the society and replace an already operating agent unit. Its initial reputation level, determined on the basis of knowledge of one of the agents will be set at 0.5 for the ta21 task. Such a level should lead to its immediate inclusion in society operation and its assignment to the business process. The results of the research have been presented in Fig. 5.

The stage marked in the figure is the time when the agent appeared in the society. In the case when the agent performs its tasks well, the results of the society's operation are good, and the process is performed correctly. In the second case, the lack of the mechanism of ostracism causes the agent to affect the correctness of the society's operation and prevent its correct operation. The inclusion of a new agent to the society through the mechanism of rumors concerning the reputation of the agent shown in Fig. 3 has caused a change of agent performing the ta21 task and the inclusion of a new ServiceAgent04 agent. As a result, the platform recorded an increase in the reputation of the society for the tested business process.

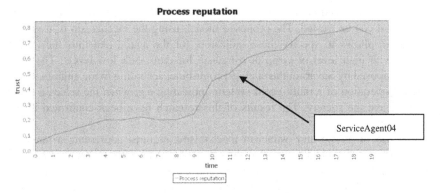

Fig. 5 An example of simulation with the use of the mechanism of gossip and ostracism

5   Conclusion

As was indicated in the paper, one of possible mechanisms contributing to definition of automorphism in multi-agent systems examined in terms of software agent societies is the mechanism of trust and reputation extended to include elements of gossip and ostracism. To illustrate the possibilities of using those mechanisms, a simulator was prepared and a range of parameters were defined that can characterise a multi-agent platform, agent societies and single agent units.

As indicated earlier in the paper, one of the solutions for the construction of a multi-agent system is the JADE platform. This platform allows for constructing multi-agent systems, however, it does not have the mechanisms associated with behavioral simulation of an agent society, supported by mechanisms of trust, reputation, gossip and ostracism. In twenty iterations that have been performed, the multi-agent system has detected an agent unit with low probability of performing the task. As a result, a new agent was introduced to the agent society, which raised the overall reputation of the process completion. The developed solution allowed to obtain positive results associated with the proper selection of purposeful tasks and enabled the improvement of indicators mentioned in this and previous articles of the author. On this basis, simulation of the operation of the society was carried out.

The result of the research is the use of model elements and proposed indicators (developed by Savarimuthu et al. [9]) in a process-oriented society of agents using the mechanism of trust and reputation. In the paper, these indicators were adopted in the developed model and used in a simulation that was carried out. As a result, the developed model allows to improve business processes, in which the agents are engaged (Fig. 5).

The results, obtained on the basis of the experiment, confirm studies of other researchers. Ken Birman indicated in the conclusions of his research [1] that "gossip is a tool, not an end in itself. It should be used selectively, in settings where gossip is the best choice". In the proposed solution the gossip and ostracism mechanisms complement the model proposed in previous research. Olajide

Olorunleke and Gordon McCalla [7] indicate in their experiments that gossip is useful, if it can be trusted. The proposed model, using the mechanism of trust and reputation, allows to specify these indicators for the agent, and thus the use of entities with high level of gosip or ostracism for units with low levels. The indicators proposed by Savarimuthu et al. [9] contribute, according to the authors, to the efficient operation of a multi-agent system, indicating e.g. when the selected agent should leave the society. The results of this research have been confirmed in the experiment.

The use of gossip and ostracism mechanisms requires addressing a range of issues connected with defining the algorithm of gossip propagation in a system, communication protocol of system entity and style of operation of the communication protocol. In the case of a society of task-oriented agents that use proposed mechanisms of trust and reputation, the dedicated mechanism of ostracism is punitive ostracism connected with the function of the society's operation. In the case of normative multi-agent systems, a better solution would be defensive ostracism that would result from fear of punishment to an agent for violating norms defined in a society of agents.

The concept of using the mechanisms of gossip and ostracism in a society of agents employing trust and reputation as elements of control and monitoring of its activity, as proposed in the paper, has a range of advantages. They include:

- Possibility of using individual agents' knowledge about the operation of other units. Such knowledge may come from information possessed by an agent as a result of its presence in another society. In the adopted model, reputation has a local character, which refers to a specific multi-agent platform. In the case of agents moving between platforms, knowledge defined in the form of gossip can support the operation of a society.
- Shortening the time during which an agent will negatively affect a specific society and its operations.
- Speeding up the moment when an agent ceases being part of a given multi-agent platform. As a result, it releases its resources, contributing to a general improvement of the performance of the operation of the platform and agents.
- Exclusion of the agent from the performance of the task by the society using the mechanism of ostracism.
- Speeding up the moment of its exclusion through the defined mechanism of gossip that was propagated by an agent and that was subject to verification during the performance of the process.
- Elimination of the agent from the platform as a result of ostracism towards its actions. Implemented only if the agent does not take any action within a specified time.

The conducted experiments concerned the operation of agents as part of a single multi-agent platform. Further research in this area should concern propagation of knowledge possessed by agents and a multi-agent platform to other multi-agent platforms.

References

1. Birman, K.: The promise, and limitations, of gossip protocols. ACM SIGOPS Oper. Syst. Rev. **41**(5), 8–13 (2007)
2. Demers, A., Greene, D., Hauser, C., Irish, W., Larson, J., Shenker, S., Sturgis, H., Swinehart, D., Terry, D.: Epidemic algorithms for replicated database maintenance. In: Proceedings of the sixth annual ACM Symposium on Principles of distributed computing, pp. 1–12, Vancouver, Canada (1987)
3. Fernandess, Y., Fernández, A., Monod, M.: A generic theoretical framework for modeling gossip-based algorithms. ACM SIGOPS Oper. Syst. Rev. **41**(5):19–27 (2007)
4. Heylighen, F., Gershenson, C.: The meaning of self-organization in computing. In: IEEE Intelligent Systems, section Trends and Controversies—Self-organization and Information Systems, 18:4, pp. 72–75, July/August (2003)
5. Kermarrec, A.M., Steen, M.: Gossiping in distributed systems. ACM SIGOPS Oper. Syst. Rev. **41**(5):2–7 (2007)
6. Klement, M., Żytniewski, M.: Metodyczne aspekty modelowania zaufania i reputacji w społecznościach agentów programowych dla potrzeb wspomagania procesów biznesowych organizacji. In: Gołuchowski, J., Frączkiewicz-Wronka, A. (eds.) Technologie wiedzy w zarządzaniu organizacją, Zeszyty Naukowe Uniwersytetu Ekonomicznego w Katowicach 243, pp. 207–226, Katowice, Poland (2015)
7. Olorunleke, O., McCalla, G.: Overcoming agent Delusi. In: Proceedings of the Second International Joint Conference on Autonomous Agents and Multiagent Systems AAMAS '03, pp. 1086–1087, Melbourne, Australia (2003)
8. Rebecca, S.B.: Some psychological mechanisms operative in gossip. Soc. Forces **34**(3):262–267 (1956)
9. Savarimuthu, S., Purvis, M., Purvis, M., Savarimuthu, B.T.R.: Gossip-based self-organising agent societies and the impact of false gossip. Minds Mach **23**(4):419–441 (2013)
10. Sommerfeld, R.D., Krambeck, H.J., Semmann, D., Milinski, M.: Gossip as an alternative for direct observation in games of indirect reciprocity. Proc. Nat. Acad. Sci. U.S.A. **104** (44):17435–17440 (2007)
11. Thomson, R.: The origins of ostracism: a synthesis. J. Hellenic Stud. **95**:243–244, The Society for the Promotion of Hellenic Studies, Copenhagen, Danish (1972)
12. Williams, K.: Ostracism—the power of silence. In: Salovey, P. (ed.) Emotions and Social Behavior series. Guilford Press (2001)
13. Żytniewski, M.: Application of the software agents society in the knowledge management system life cycle. In: Pańkowska, M., Stanek, S., Sroka, H. (eds.) Cognition and Creativity Support Systems, Publishing House of the University of Economics in Katowice, pp. 191–201. Katowice, Poland (2013)
14. Żytniewski, M., Sołtysik, A., Sołtysik-Piorunkiewicz, A., Kopka, B.: Modeling of software agents' societies in knowledge-based organizations. The results of the study. In: Proceedings of the 2015 Federated Conference on Computer Science and Information Systems, IEEE, September (2015)
15. Żytniewski, M.: Wprowadzenie do teorii społeczności agentów programowych oraz ich zastosowania w organizacjach opartych na wiedzy. In: Żytniewski, M. (ed.) Technologie agentowe w organizacjach opartych na wiedzy, Wydawnictwo Naukowe Uniwersytetu Ekonomicznego w Katowicach. Katowice, Poland (2015)
16. Żytniewski, M.: Modelowanie systemów agentowych wspomagających organizacje oparte na wiedzy. In: Żytniewski, M. (ed.) Technologie agentowe w organizacjach opartych na wiedzy, Wydawnictwo Naukowe Uniwersytetu Ekonomicznego w Katowicach. Katowice, Poland (2015)
17. Żytniewski, M., Klement, M.: Trust in software agent societies. Online J. Appl. Knowl. Manag. Publ. Int. Inst. Appl. Knowl. Manage. **3**(1), 93–101 (2015)

18. Żytniewski, M.: Integration of knowledge management systems and business processes using multi-agent systems. In: Alqithami, S., Hexmoor, H. (eds.). Int. J. Comput. Intell. Stud. **5** (2016) (Special Issue on: "Networked Agents of Complex Online Organisations")

19. Żytniewski, M., Kopka, B.: Proposal for using analysis of software agents usability in organisations. In: Alqithami, S., Hexmoor, H. (eds.). Int. J. Comput. Intell. Stud. **5** (2016) (special issue on: "Networked Agents of Complex Online Organisations")

20. Żytniewski, M.: Comparison of methodologies for agents' software society modeling processes in support for the needs of a knowledge-based organization. In: Kiełtyka, L., Niedbał, R. (eds.) Wybrane zastosowania technologii informacyjnych zarządzania w organizacjach, Publishing House of University of Technology in Częstochowa, vol. 296, pp. 15–26. Częstochowa, Poland (2015)

21. Żytniewski, M., Klement, M.: Analiza porównawcza wybranych platform wieloagentowych. In: Kisielnicki, J., Chmielarz, W., Parys, T. (eds.) Informatyka 2 Przyszłości, Wydawnictwo Naukowe Wydziału Zarządzania Uniwersytetu Warszawskiego, pp. 88–100, Warszawa, Poland (2014)

22. Żytniewski, M.: Mechanizmy reprezentacji wiedzy w hybrydowych systemach wieloagentowych. In: Filipczyk, B., Gołuchowski, J. (eds.) Wiedza i komunikacja w innowacyjnych organizacjach. Systemy ekspertowe—wczoraj, dzis i jutro, pp. 126–133, AE Katowice, Katowice, Poland (2010)

Must-Opt Imperatives and Other Stories Make Passengers of Low Cost Carriers' Feel Put-upon: User Perceptions of Compliance with EU Legislation

Chris Barry, Mairéad Hogan and Ann M. Torres

1 Introduction

It is reasonable to expect Web technologies to be employed in a manner that enhances the user experience, allowing them to engage in a satisfying and productive interaction. Yet, all is not well with this expectation. To encourage the purchase of optional ancillary services, some low cost carriers (LCCs) are using a new approach of forced choice, which is referred to in this paper as a 'must-opt' selection. Although this approach is clever in persuading consumers to avail of ancillary services, it contravenes European legislation that governs airline ticket sales and is designed to protect consumers from unfair practices. The LCC model, consumer protection regulation, and the 'grey' Web practices of the 'must-opt' are outlined before exploring users' perceptions of two LCCs' compliance with the relevant European legislation.

A prior version of this paper has been published in the ISD2016 Proceedings (http://aisel.aisnet.org/isd2014/proceedings2016).

C. Barry (✉) · M. Hogan · A.M. Torres
National University of Ireland Galway, Galway, Ireland
e-mail: chris.barry@nuigalway.ie

M. Hogan
e-mail: mairead.hogan@nuigalway.ie

A.M. Torres
e-mail: ann.torres@nuigalway.ie

© Springer International Publishing Switzerland 2017
J. Gołuchowski et al. (eds.), *Complexity in Information Systems Development*,
Lecture Notes in Information Systems and Organisation 22,
DOI 10.1007/978-3-319-52593-8_10

2 LCCs' Influence and Ancillary Revenues

Succinctly described, the LCC model offers 'no-frills passenger service within a point-to-point network of short and medium-haul routes that serve secondary airports, using a highly productive homogenous fleet' [1]. The model is designed to have a competitive cost advantage over full service carriers (FSCs) by reducing unit costs while simultaneously increasing output and productivity [2].

An LCC entering a market results in a significant decrease in airfares, a notable increase in passenger volumes and an expansion of the catchment area [2, 3]. Consequently, the LCCs' impact on the airline industry has been profound; they have altered the industry structure and shifted the basis for competition, efficiency and consumers' expectations [4–7].

LCCs have garnered higher market shares on many domestic and intra-European routes [8]. To compete effectively, some European FSCs have created low-cost subsidiaries (e.g., KLM's Transavia, Lufthansa's Germanwings, and SAS's Snowflakes). Even those FSCs without low-cost subsidiaries have adopted certain LCC management practices, such as increasing plane and crew utilisation, eliminating business class, and introducing paid ancillaries.

'Once largely limited to low fare airlines, ancillary revenue is now a priority for many airlines worldwide' [9]. Among the airlines currently accruing the highest levels of ancillary profits are global FSCs (e.g., United, Delta, American, Qantas, Air France and Korean Air) [9]. Baggage fees were once the largest single source of ancillary revenue [10, 11]. Now, various ancillary services (e.g., priority boarding, meals, and seat selection), co-branded credit cards, and third party vendor commissions (i.e., insurance, car and hotel bookings) contribute significantly to airline profitability. Unbundling services boosts profit margins, whereby, 'the most aggressive airlines generate more than 20% of their revenue' through the sale of ancillary services [9]. In 2015, ancillary revenues accounted for 25% of Ryanair's total revenue [12] and 13% of Aer Lingus's total revenue [13]. Global ancillary revenue in the airline industry has more than doubled from $22.6 billion in 2010 to $59.2 billion in 2015 [11, 14].

3 Opt-In, Opt-Out or Must-Opt

The LCCs' adoption of technology, in areas such as electronic ticketing and dynamic pricing, has become essential in offering consumers efficient flight options. Despite these advances, a number of LCCs use their information systems in a conflicting manner, especially when selling ancillary services. The Websites smoothly engage and facilitate customers through the self-service process to commit users to purchase tickets. However, once users move beyond the 'committal' point (i.e., after selecting where and when to travel and receiving an initial quote) and ancillary services are introduced, the Websites appear more opaque. Research on this phenomenon has found significant disquiet amongst users [15]. Ancillary services are normally offered as optional through some type of opt-in or opt-out mechanism. EU regulation 1008/2008 requires optional extras to be accepted by the consumer on an opt-in basis only [16]. Yet, the legislation does not define what is meant by the term 'opt-in'. The definitions of opt-in and opt-out vary depending on the source. According to the Oxford English Dictionary (OED) [17], the term opt-in means, 'to choose to participate in something', whereas opt-out means 'to choose not to participate in something.' According to Wiktionary [18], opt-in means 'Of a selection, the property of having to choose explicitly to join or permit something; a decision having the default option being exclusion or avoidance.' Whereas, opt-out means 'Of a selection, the property of having to choose explicitly to avoid or forbid something; a decision having the default option being inclusion or permission.' However, the airlines are using a new approach, referred to in this paper as a 'must-opt' selection. This format requires users to explicitly accept or reject the service before continuing with the interaction.

Based on the OED definition, the airlines' must-opt options are both opt-in and opt-out, as the user must choose explicitly whether to participate [17]. Yet, based on the second part of the Wicktionary definition, the must-opt is neither opt-in nor opt-out, as the default option is to prevent the user from continuing until they choose to accept or refuse the option [18]. The airlines are following the OED's definition of opt-in for their optional extras, except they also conform to its definition of opt-out. Airlines would presumably argue, as they comply with the definition of opt-in, they comply with the legislation.

While little research examines the effect of opt-in versus opt-out in retail sales, research in other fields demonstrate their impact on users' decisions. Johnson and Goldstein's study [19] suggest changing the default option for organ donation to opt-out increased organ donation. Madrian and Shea [20] find people are more likely to proceed with the default option, due to inertia and a belief the selected option is recommended. McKenzie et al. [21] concur with this finding, which suggests airline consumers are more likely to purchase ancillary services if they are presented as an opt-out. Consequently, airlines prefer to use opt-out as the default. As EU regulation prevents this approach, airlines needed another way to increase the likelihood of choosing an option.

The airlines looked to instances where the customer must choose an option before proceeding. This forced choice is generally associated with accepting terms and conditions (e.g., licence agreements for a software installation). Airlines have adapted this forced choice by presenting consumers with a must-opt. While the airlines cannot use the implicit recommendation inherent in an opt-out, some use an explicit recommendation to encourage purchase.

4   EU Regulation

The European Commission co-ordinated the airline ticket selling investigation under the auspices of Consumer Protection Co-operation Regulation, which came into force in 2006 [22]. The report identified the most common unfair practices related to price indications, limited availability of special offers, and contract terms. The Commissioner directed airlines to give the total price, including taxes and booking/credit card fees, in the first advertised price on a Website. Other unfair practices include the mandatory purchase of insurance, or presenting optional services (e.g., insurance, priority boarding, seat selection) as opt-outs.

This report led to the introduction of EU regulations to prevent airlines from pursuing these practices [16]. Failure to comply may result in legal action or

Table 1 Research objective and research questions

Research Objective (RO): to identify user perceptions on whether LCC airlines comply with EU regulations in respect of price transparency and fair B2C commercial practices.

Source	Provision	Research question
Regulation 1008/2008 Article 23	'The final price to be paid shall at all times be indicated and shall include the applicable air fare or air rate as well as all applicable taxes, and charges, surcharges and fees which are unavoidable and foreseeable at the time of publication'	RQ1: Does the first displayed price include all unavoidable and foreseeable charges? Justification: Since the final price, including charges must be visible at all times; the first price must also include this information
Regulation 1008/2008 Article 23	'The final price to be paid shall at all times be indicated and shall include the applicable air fare or air rate as well as all applicable taxes, and charges, surcharges and fees …'	RQ2: Is the total price clear at all times during the booking process?
Regulation 1008/2008 Article 23	'In addition to the indication of the final price, at least the following shall be specified: (a) air fare or air rate; (b) taxes; (c) airport charges; and (d) other charges, surcharges or fees, such as those related to security or fuel …'	RQ3: Is a detailed breakdown of all price elements specified?
Regulation 1008/2008 Article 23	'Optional price supplements shall be communicated in a clear, transparent and unambiguous way at the start of any booking process'	RQ4: Are all optional extras communicated in a clear, transparent and unambiguous way, at the start of the booking process?
Regulation 1008/2008 Article 23	'… their acceptance (optional prices) by the customer shall be on an 'opt-in' basis'	RQ5: Are all optional extras presented as opt-ins?
Directive 2005/29/EC	Derived from Articles 8 on 'Aggressive commercial practices' and Article 9 on the 'Use of harassment, coercion and undue influence'	RQ6: Do users feel harassed or coerced into choosing optional extras?

closure, as well as being 'named and shamed' for failing to comply with EU law. European legislation governing airline price information is found in Article 23(1) of Regulation 1008/2008 [16] and Articles 5–7 of the Unfair Commercial Practices Directive [23]. The first of these, Article 23(1), applies only to Airlines operating in the EU, whereas the second, Articles 5–7, applies to all business to consumer transactions within the EU [23]. These articles provide specific guidance on acceptable pricing and consumer protection practices in the airline sector and e-commerce generally.

The main research objective was to identify user perceptions on whether LCC airlines comply with EU regulations in respect of price transparency and fair B2C commercial practices [24]. The provisions of the articles relevant to the main research objective are laid out in Table 1, along with the research gap and the articles from the regulations that form the basis for this study's research questions.

5 Research Methodology

Verbal protocols involve a typical user thinking out loud while carrying out representative tasks on a system. As they carry out tasks, the participant explains what they are doing and why [25]. This verbalisation helps the evaluator to understand the user's attitudes towards the system and to identify problematic aspects [26]. A key strength of this technique is showing users what their doing and why, while they are doing it [27].

In this study, verbal protocols were used to determine users' opinions as to whether the airlines comply with EU legislation governing the sale of flights via the Internet. Twenty typical users of LCC Websites participated in a series of verbal protocol evaluations. While carrying out the tasks, the participant was prompted to talk aloud and consider the Website's level of adherence to the legislation. Participants were briefed on the verbal protocol approach and were encouraged to think of the session as a structured discussion, where the LCC websites were being examined, not the participant.

Each participant was shown a summary of the legislation. They were then asked to define what they understood by the term opt-in. As the legislation requires consumers only accept optional extras on an opt-in basis [16], it was important to know participants' understanding of the term. This approach permitted the researcher to remind participants of their definition, if necessary.

The participants were required to find and book a flight for each airline. The order of the airlines was counterbalanced to ensure it would not bias the results. As the participants worked through the tasks, they were reminded of the legislations' requirements. When they encountered prices, optional extras or difficulties, they were asked whether they believed the airline complied with legislation. They were also asked why they believed the airline designed the website as they did. To ensure the prompts did not distort or invalidate the user's dialogue, the prompts were developed in advance and used at appropriate times [28]. The participants completed a short questionnaire after each airline protocol to collect their opinion on each airline's legislative compliance.

5.1 RQ1: Does the First Displayed Price Include All Unavoidable and Foreseeable Charges?

RQ1: Ryanair The first quoted price on Ryanair's website was highlighted in a banner for the first leg of the journey. For a return journey, the second leg was displayed further down the page. The prices did not include taxes, charges or other unavoidable costs. Below each banner, the flight cost was broken down in a box. When participants were asked whether the first quoted price included all unavoidable and foreseeable charges, the responses fell into three categories: (1) an instant positive response followed by caution: 'Yes, I'm hoping everything is included'; (2) a qualified positive response: 'from Ryanair, I don't know. It looks like it' and (3) an instant negative response because of a priori views about Ryanair: 'No. I'm sceptical of Ryanair'.

Views differed on whether Ryanair complied with legislation requiring the first quoted price to include all unavoidable and foreseeable charges. While the first price in the banner included no additional charges, such as taxes or check-in charges, the box below the banner displayed a price breakdown. Although some respondents needed the box's details pointed out to them, most were happy the total price included all unavoidable and foreseeable charges: '*Yes. It's showing me online check-in, taxes and charges, flight cost*'. Nonetheless, many were still sceptical: '*the first price—No, but the total price—Yes. Except, I've had experience of Ryanair before*'.

Others believed the price did not include all unavoidable and foreseeable charges. Many participants believed the handling charge was unavoidable, even though MasterCard Prepaid did not invoke a charge. As one participant stated: '*the administration charge is technically avoidable. [Ryanair is] compliant but it's sneaky*'. Despite the cynicism, many participants thought Ryanair's price break-down was clearer than Aer Lingus' price breakdown. Nine participants believed Ryanair was compliant, while eleven did not.

RQ1: Aer Lingus Aer Lingus displayed prices in a similar way to Ryanair. The price for the date selected was presented in a banner with the detailed price displayed below. In contrast to Ryanair, the initial responses to Aer Lingus were more positive, falling into the following categories: (1) an instant positive response followed by a change of mind later in the process: '*Yes. I am going to trust them. From experience, I like them a bit more*' (when this user saw the handling charge added on at a later stage in the process, they said: '*If I hadn't read the footer I'd be surprised. It [the handling charge] is unavoidable.*'; (2) a qualified positive response where having to scroll to see a price inclusive of taxes and charges was considered acceptable: '*Yes, if it's on one screen*', '*It would be better if it were all* *up on top since they know the charges'*; and (3) an instant negative response because of a priori views about Aer Lingus: *'The price should [contain all unavoidable charges]. As a regular user, I expect other foreseeable charges. A first time user might have different expectations'*.

As the process continued, no participants believed Aer Lingus was compliant. The handling fee was not added until the third 'total price' was displayed. However, a footnote stated a handling charge would be added, unless the flight originated in the USA or Canada or was paid by Visa Electron. This footnote was in small, light grey text and positioned below the continue button of each page. The text position and pale colour meant many participants did not see the statement: *'I don't think anyone will read the extra information at the bottom of the screen'* and were surprised when it was added at a later stage: *'Handling fee? I thought that was included. It's a surprise'*. Most participants believed the handling fee was unavoidable and should have been included in the initial price. However, some did believe it could be avoided, but were unsure how. The footnote stated there was no handling charge, if the customer paid by Visa Electron. This was not the case, as one participant stated: *'Visa Electron is not removing the handling charge'*. The only way to avoid the handling charge was by booking a flight that originated in the US or Canada. For flights originating in Ireland, the handling fee was unavoidable, but was not included until the third time the 'total price' was displayed.

The participants were more negatively disposed to Ryanair initially. Although, when the booking process was complete, participants were more negatively disposed towards Aer Lingus because of the way it managed the handling charge. Many participants were surprised Ryanair was more compliant than Aer Lingus, particularly given their initial negative disposition towards Ryanair.

The participants believed Aer Lingus displayed the prices in a way to mislead people, as one participant stated: *'to make the price look more attractive than it is'*. Even though this approach resulted in annoyed customers when the 'total price' was revealed, participants believed Aer Lingus used this design deliberately to lure customers further into the booking process. As one participant stated: *'[it is part of] the psychology of presenting the lowest figure. You are committed to the process. They know you have wasted time and probably won't go elsewhere'*. Aer Lingus appears less concerned with customer satisfaction and more concerned with enticing the customer in; as one respondent put it: *'puffery and gimmicky'* on the presumption the customer would continue once they commit sufficient time to the booking process.

5.2 RQ2: Is the Total Price Clear at All Times During the Booking Process?

RQ2: Ryanair This question was asked at several points during the interaction. The initial responses for Ryanair were positive. Some participants were unhappy the total for each leg was displayed and the user then had to scroll down to see the full total. Once this initial page is passed, Ryanair provide a running total at the side of the screen. Participants were satisfied with this easy-to-read approach.

As the process continued, it became less clear because the customer must first select the number of bags to see the cost. At the payment screen, it became 'fuzzy' because an administration charge is added once the customer selects a payment type other than MasterCard Prepaid, which was not obvious until it was pointed out to participants.

RQ2: Aer Lingus There was a mixture of initial responses for Aer Lingus, with some participants initially believing it was compliant, but changing their mind once the handling fee was added. Others believed it was non-compliant from the beginning as it was necessary to scroll down to see the total price on the first screen. There was significant dissatisfaction with the way the handling fee was administered, with most participants not noticing the statement regarding the fee, which was positioned low on the screen and in grey text. One participant stated the text was greyed *'so that you don't read it'*.

Another irritant was the booking process for two passengers; the price displayed was the total price per passenger until the third time it was presented, when it doubled and the handling charge was added. Most participants believed the original 'total price' was for two people and were shocked when the price doubled. At this point, none believed Aer Lingus complied with the requirement for the total price to be clear throughout the booking process. Most participants believed the price displayed was deliberately confusing to make customers feel committed and conclude the purchase.

5.3   *RQ3: Is a Detailed Breakdown of All Price Elements Specified?*

RQ3: Ryanair Ryanair was more transparent than Aer Lingus in outlining taxes and charges. They are listed as part of the total and when the user clicks on the Taxes/Fees hyperlink, they are broken down into 'taxes and fees' and 'aviation insurance/PRM levy'. However, most participants believed this approach did not add clarity to their understanding of what the different components were. As one participant stated: 'I have no idea what the PRM levy or the Aviation Insurance is'. While the majority of participants believed Ryanair was non-compliant, they were more concerned a true total for taxes and charges was displayed.

RQ3: Aer Lingus Aer Lingus listed taxes and charges as a hyperlink in the table displaying the price, where additional information about taxes and charges may be found. However, it does not state the charges. One participant stated: *'it's just telling me I have to pay charges'*. Another participant believed Aer Lingus were hiding the charges, stating: *'(they are) trying to blindside you and not disclose what* *charges apply to you*', while another believed Aer Lingus was simply being lazy by including details of charges for US flights in the pop-up window.

Although a breakdown was not important to participants, they did want to see a total for all taxes and charges. Aer Lingus not including the handling charge within this total was unexpected and unacceptable to most participants. Many participants initially believed the handling charge was included in the taxes and charges and were surprised when it was added at a later stage. Some participants believed the handling charge was added separately to fool the customer into believing the taxes and charges were less than they were: '*[it is done] so that the taxes and charges don't appear large*'.

5.4 RQ4: Are All Optional Extras Communicated in a Clear, Transparent and Unambiguous Way, at the Start of the Booking Process?

RQ4: Ryanair Unanimously, participants believed neither airline complied with the provision that all optional extras be communicated in a clear, transparent and unambiguous way at the beginning of the booking process. Ryanair's flight selection screen displayed the message '*Optional charges such as administration and checked baggage fees are not included*' beneath the total flight cost. Despite its proximity to the flight cost most did not see the message and found it unhelpful in clarifying specific optional charges. Nine other optional charges or services were not mentioned. A 'detail' link invokes a pop-up window where fees (i.e., not optional extras) were explained. Some participants thought the pop-up acceptable in presenting detail of optional charges and others believed the information value was lost because the content was extensive and superfluous.

RQ4: Aer Lingus With Aer Lingus, no indication other than a handling charge was displayed on the initial selection page, which was difficult to see because it was at the bottom of the screen and greyed out. On the next screen many participants expressed severe annoyance the total flight price had changed with the addition of an unexplained handling fee. Many believed optional extras were introduced incrementally so the price would change gradually and users would be less likely to back out of the process. In this regard, participants believed such design to be deliberate.

5.5 RQ5: Are All Optional Extras Presented as Opt-Ins?

RQ5: Ryanair This question dealt with the 'services' pages and generated enormous confusion, ignominy and anger. Ten decisions on optional extras must be negotiated. Seven of them were 'must-opt' because they forced a choice before proceeding. The mechanism for enforcing a must-opt is a pop-up error window when the 'Continue' button is pressed detailing the option has not been selected. Barring Terms and Conditions, no indication is given that any must-opt demands an interaction on the part of the user. The remaining three options (i.e., baggage, sport equipment and special assistance) are opt-in and can be by-passed without any required interaction.

Participants were unclear and wary about the nature of options and spent some time reading them to avoid choosing them. Participants were asked what they understood by an opt-in and opt-out selection. Nearly all defined opt-in and opt-out in a way consistent with the definitions in the literature review. There was clarity and satisfaction amongst participants that the bag option was actually opt-in. Priority boarding presents the first must-opt decision point. Five participants believed it was an opt-in decision, but as their comment was queried, they reversed their initial view. More believed it was either opt-out or at least forcing them to make a decision, coming close to the definition that opting-out was explicitly choosing to avoid or forbid something. Others expressed confusion about the exact nature of the interaction with comments like: *'it's making me choose'*; *'it's forcing me to make an option'*; and *'it's making me read through it.'* Participants were uneasy about the design of priority boarding and the predominant view was the option neither complied with the letter nor the spirit of the law.

Travel insurance caused considerable resentment and anger amongst participants. The design is at first curious and ultimately devious. The user is invited to buy travel insurance using neither Yes/No radio buttons nor a check box, but rather a drop down list with the default option to 'Please select a country of residence', a supposition the service has been chosen. Users were drawn to the alphabetically ordered drop down list to look for an avoidance mechanism. About half way down, between Latvia and Lithuania was a 'country' called 'No Travel Insurance Required.' Most did not notice the line beneath the drop down list that informed participants that 'If you do not wish to buy insurance select No Travel Insurance in the drop down menu.' Participants described this feature in trenchant terms, commenting: *'it's buried into the drop down list'*; and *'this should be illegal!'* It drew a torrent of adjectives such as: underhand, sneaky, aggressive, extremely dodgy, tricky and deceptive. Participants wholly agreed the design was intentional and was not legislatively compliant.

Regarding the SMS optional extra, participants were presented with two unchecked radio buttons with no indication it bears a cost. Several participants found this feature useful; none believed the charge appropriate. While it was a must-opt, several believed it was opt-in, some believed it was opt-out and others believed it was neither opt-in nor opt-out. Participants did not express strong emotions on compliance, although more believed it was not compliant.

The approved Ryanair cabin bag elicited substantial emotional reaction. Participants believed it was marketing gimmickry and, at worst, pushy marketing. What annoyed them most was that it delayed the booking process. A few believed it gave the impression it was an official cabin bag.

Regarding sports equipment and special assistance, general comments were benign and participants quickly realised they were genuine opt-in decisions. Ryanair use a must-opt check box for Terms and Conditions, a widely employed convention in transactional activity and for registration purposes.

There was surprise and annoyance when a pop-up page reminded the participant to make selections on the must-opt decisions that were overlooked: '*it is making me read all the options*'; '*[it] didn't say it was compulsory to answer travel insurance*'; '*you should not be forced to decide.*' A further inconvenience for participants was that going back a screen involved re-entering all decisions except the passenger's name. On declining the must-opts and choosing 'Continue' on the main services page, a pop-up must-opt decision for travel insurance appeared. Participants were implored to '*Wait!*' asking if they were prepared to risk not buying travel insurance. Most believed the reminder was pushy and did not conform to an opt-in decision.

The second service page is devoted to selling a Hertz rent-a-car. The first reaction of participants to the first named traveller pre-selected as the main driver and one of three cars highlighted, was it had been chosen. One commented '*Oh my God—it looks like they rented a car for me!*' At this point participants were irritated and exhausted, one wearily commenting there's '*a lot of reading to make sure I don't miss something.*' Participants believed such heavy selling was inappropriate in the middle of the process of booking a flight and that it would be better placed after the flights were reserved. The general view was that it was non-compliant.

RQ5: Aer Lingus Users navigate through four screens that deal with the booking process up to card payment. Fourteen decisions on optional extras must be negotiated. Of these, three are must-opt, one opt-out and the remainder opt-in. Participants defined the concepts of out-in and opt-out consistently with the definitions in the literature review.

Participants agreed the 'flex fare' optional extra (i.e., offering free date changes and lounge access) on the first page was opt-in. Some believed it slightly pushy as it was embedded within the booking process. On the next page the participant encountered another invitation to choose a flex fare for each leg of the journey. It was not immediately obvious to participants that it was the same flex option from the previous screen. Many were irritated they had to consider the option twice. Aer Lingus also used a must-opt check box for Terms and Conditions to confirm they were read.

Participants agreed SMS was opt-in and there were no significant issues. They believed it consistent with their definition at the outset. One reported '*it's opt-in, that's the default. You have to knowingly choose it.*' It was also positively noted by several the SMS price was shown before it was selected. Participants believed the mailing list opt-out option was unclear and 'tricky' since previous optional extras were opt-in. Several expressed dissatisfaction with the way it was designed, immediately beneath the SMS option in lighter grey text, and header-less. Three opt-in options (i.e., frequent flyer, special assistance and voucher submission) followed, that participants easily by-passed, and drew no comment.

Most participants believed baggage was compliant, as it was a clear opt-in decision. However, to see the price participants had to click a drop down list. It was mentioned this approach was preferable to Ryanair's design, whereby the baggage had to be selected before a price appeared.

Travel insurance was presented in a confusing manner and invoked annoyance amongst participants. The amount appeared pre-selected in a right hand column, but was not included in the total price, further down the page. Two Yes/No options appeared with un-checked radio buttons; thus it was a must-opt. There was disagreement on how to describe the decision. While few believed it was opt-in, participants were fairly evenly split as to whether it was opt-out, neither, or unsure. Many changed their minds during the interaction. The confusion was more exaggerated with Aer Lingus than Ryanair because the cost was pre-selected alongside other charges.

Participants dealt with opt-in lounge access speedily. Parking was presented as a must-opt selection, displayed identically to the travel insurance. Thus, the cost was shown in the right hand side of the page as if it had already been selected. By the time participants dealt with the task of contemplating parking, many were weary. Once again they were required to read text carefully to avoid a charge and confirm they did not want it. Some believed it was presumptuous to ask them about parking. When asked why parking and travel insurance were designed in this way, one participant cynically observed '*they prioritise what they want to sell and then choose different technology [to sell it].*' On compliance, participants were of two views on Aer Lingus's travel insurance and parking options: they either wholly contravened the legislation or were designed to get around it.

5.6   RQ6: Do Users Feel Harassed or Coerced into Choosing Optional Extras?

RQ6: Ryanair Decisions on baggage, SMS, sports equipment and special assistance presented no problems. There were a few concerns over priority boarding, namely the way in which it was phrased and presented. The option to purchase a Ryanair approved cabin bag was more problematic. Participants believed it was inappropriate and described it as pushy and annoying. Some believed there was an element of coercion to ensure their bag would be acceptable to Ryanair. Regarding travel insurance, where the option to avoid insurance was half way down a drop down list, many were annoyed and believed it was unacceptable design; several believed it was deceptive. On declining travel insurance, the pop-up reminder was widely cited by participants as aggressive and coercive. Strong emotions were also evinced on the design and colour choices that suggested danger for declining travel insurance. Participants agreed Ryanair were non-compliant on this question.

Most participants reacted to the Hertz rental page with distaste. Views ranged from pushy but compliant, to aggressive and non-compliant, to extremely aggressive and non-compliant. Comments included: '*a little bit aggressive and harassing at this stage*'; '*this is aggressive selling. It's moving towards coercion. [They] pre-*

selected my name!'; '*[it's] aggressive marketing. It's definitely non-compliant'*; and '*it's pushy, aggressive, in my face.'*

RQ6: Aer Lingus Generally, participants believed Aer Lingus was more compliant than Ryanair. Most believed Aer Lingus's site was not usually harassing or coercive. However, there were areas of concern. The first flex fare was thought acceptable, but when it appeared a second time, most thought it was pushing the boundaries of compliance; '*(it's) pushy. A bit aggressive, borderline compliant'*; '*it's designed to be close to the border of coercive. Still it is pushing it.'* Some believed the must-opts (e.g., insurance and parking) with Aer Lingus were borderline compliant, while others believed they were outright non-compliant. The pre-selected nature of the cost was thought to be pushy: '*they are hoping people will take it. It's a little aggressive'*; '*it's designed to just meet the terms of the legislation.'* Others complained about the must-opt design of these optional extras, one declared: '*it's aggressive. I shouldn't have to tell them I don't want it.'*

6 Discussion and Conclusions

Respondents expressed differing initial views on whether they believed the first quoted price on Ryanair included all unavoidable and foreseeable charges. A few believed optimistically it did; but most had varying degrees of scepticism. After some time spent perusing the screen, participants began to take in more of the breakdown details displayed. Despite the reasonable granularity of taxes and charges, more than half of the participants did not believe Ryanair was in compliance with the legislation requiring the first quoted price to include all unavoidable and foreseeable charges. A positive disposition was expressed towards Aer Lingus in respect of this price compliance; but, as the interaction continued, this predisposition of greater trust evaporated. In summary, Ryanair exceeded users' low expectations and Aer Lingus fell short of matching expectations.

Regarding consistent clarity of the total price, including all charges and selected extras, any early positivity melted away as participants navigated the booking process. Efforts to increase price visibility were negated by tricky navigation and irksome features. Participants seemed unbelieving even when early pages offered some clarity.

Participants were also sceptical as to whether the airlines complied with the EU legislation governing the requirement for the components of taxes and charges to be clear throughout the quote and booking process. The airlines displayed composite taxes and charges, but did not provide a clear and comprehensive breakdown of charges. However, participants displayed little enthusiasm for the minutia of taxes and charges. While neither airline particularly impressed participants, Ryanair communicated greater transparency than Aer Lingus who irritated with murky design features.

No participant believed the airlines complied with the requirement to communicate all optional extras in a clear, transparent and unambiguous way at the beginning of the booking process. Apparently the airlines ignored this EU legislative obligation [23] governing airlines' price information. Perhaps they do so to maintain the low cost perception as long as possible in the booking process. By delaying the user from realising the true flight cost until late in the interaction, it makes comparisons across airlines more difficult. Therefore users are less likely to quit the process, having invested heavily in time and emotional capital.

The findings are conclusive that all optional extras are not presented as opt-in decisions to users. A key finding in this study is that a novel device, a 'must-opt', has been deployed, perhaps to circumvent the legislation. Must-opt decisions are both opt-in and opt-out and are also neither opt-in nor opt-out. Strictly speaking a user does not need to read opt-in or opt-out decisions, which is most decidedly not the case with a must-opt decision. A user may not proceed to the next screen without making a selection. Users failure to do so means they enter an endless loop until they comply. Hence, users are forced to engage intellectually and mechanically with the Web page. Must-opt decisions apply pressure on users to take seriously the option placed before them. In some cases this approach is justifiable, such as the requirement that users read terms and conditions of a 'contract'. The must-opt is clever; it does not pre-select an option for a user, so ostensibly it is not an opt-out optional extra. More importantly, it does not allow the user to lazily scroll through unwanted services seeking the continue button and an exit from the page. It presents a juncture in the workflow of the transaction where a user must select either service inclusion or exclusion.

At the outset participants were clear about their definition of opt-in. However, this understanding often changed during the verbal protocol session, sometimes shifting towards an acceptance of must-opt as a characteristic of the opt-in process. This uncertainty surrounding users' understanding of choice led to confusion, wariness and frustration. Ryanair presenting the must-opts in a variety of design forms added further confusion. The reason for this design is unclear, but poor systems development practices would be the most innocent explanation. Given LCCs' awareness of certain industry conventions, such as enforcing users to agree terms and conditions have been read, means they appreciate what a must-opt construct is designed to achieve. However, as ancillaries increase their contributions to the airlines' bottom-line, it can be expected airlines will continue to find creative ways to circumvent the legislation's ethos.

Participants' opinions were mixed as to whether the airlines were using harassment or coercion to convince them to purchase optional extras. Most issues arose around the must-opt selections. Participant views ranged from deeming them pushy to harassing and non-compliant. While insurance on both Websites is presented as a must-opt, participants responded more negatively to Ryanair's handling of the option. Re-presenting the option again after users have clearly made a decision questions the merit of their action rather than pointing out something they have overlooked. The offer of parking on Aer Lingus can be seen as similar to the offer of car hire on Ryanair in that both are tangential to the task of booking a flight. However, even though Aer Lingus use the must-opt feature for parking, the reactions of the respondents were not as negative as they were for car hire on Ryanair.

Presumably this reaction is because, although both use a must-opt selection, Aer Lingus present it in a more muted way, as an option in the flow of the purchase; Ryanair presents it as a separate window using bold black and yellow lettering.

This study set out to explore user views on whether two Irish LCCs were acting in good faith in implementing consumer protection legislation, enacted once the European Commission recognised new technologies were used to nudge consumers to behave in ways airlines preferred. While ancillaries have become an essential and growing component of airlines' revenues, it should be expected they are implemented in a way where consumers do not perceive them as a barrier to securing flights. Although regulations specifically recommend optional charges be accepted on an opt-in basis, the airlines in this study may have found a technological by-pass of the regulations—the must-opt construct. This approach and the ambiguity of the definition of opt-in and opt-out decisions allow airlines to exploit the legislation to serve their own ends. The airlines' lack of full compliance with EU legislation suggests forthcoming EEC-Net reports are likely to reiterate the same recommendations. The game of catch-up between regulation and technology continues.

References

1. Daraban, B.: The low cost carrier revolution continues: evidence from the US airline industry. J. Bus. Econ. Res. **10**, 37–44 (2012)
2. Alderighi, M., Cento, A., Nijkamp, P., Rietveld, P.: Competition in the European aviation market: the entry of low-cost airlines. J. Transp. Geogr. **24**, 223–233 (2012)
3. Mertens, D.P., Vowles, T.M.: The southwest effect—decisions and effects of low cost carriers. J. Behav. Appl. Manag. **14**, 53–69 (2012)
4. Franke, M.: Innovation: the winning formula to regain profitability in aviation? J. Air Transp. Manag. **13**, 23–30 (2007)
5. Graham, B., Vowles, T.M.: Carriers within carriers: a strategic response to low-cost airline competition. Transp. Rev. **26**, 105–126 (2006)
6. Shumsky, R.: The southwest effect, airline alliances and revenue management. J. Rev. Pricing Manag. **5**, 83–89 (2006)
7. Rubin, R.M., Joy, J.N.: Where are the airlines headed? Implications of airline industry structure and change for consumers. J. Consum. Aff. **39**, 215–228 (2005)
8. Fageda, X., Jimenez, J.L., Perdiguero, J.: Price rivalry in airline markets: a study of a successful strategy of a network carrier against a low-cost carrier. J. Transp. Geogr. **19**, 658–669 (2011)
9. Sorensen, J.: CarTrawler Review of Ancillary Revenue Results for 2012. IdeaWorksCompany.com (2013)
10. Harteveldt, H., Stark, E.: Airlines Need to Convince Passengers to Use Digital Channels to Buy Ancillary Products. Forrester Research, Cambridge, MA (2010)
11. Sorensen, J.: Planes, Cars, and Ancillary Revenues. IdeaWorksCompany.com (2011)
12. Ryanair Annual Report. https://investor.ryanair.com/wp-content/uploads/2015/07/Annual-Report-2015.pdf
13. Aer Lingus Group Plc, 2015 Q2 & H1 Results. http://corporate.aerlingus.com/media/corporateaerlinguscom/content/pdfs/H1_2015_presentation_FINAL.pdf
14. Sorensen, J.: CarTrawler Review of Ancillary Revenue Results for 2015. IdeaWorksCompany.com (2015)

15. Torres, A., Barry, C., Hogan, M.: Opaque web practices among low-cost carriers. J. Air Transp. Manag. **15**, 299–307 (2009)
16. European Union: 1008/2008. Regulation of the European Parliament and of the Council on Common Rules for the Operation of Air Services in the Community (recast) (2008)
17. The Oxford English Dictionary. http://www.oed.com/view/Entry/132049?redirectedFrom= opt+in#eid33209114
18. Wikitionary. http://en.wiktionary.org/wiki/opt-in
19. Johnson, E.J., Goldstein, D.G.: Do defaults save lives? Science **302**, 1338–1339 (2003)
20. Madrian, B.C., Shea, D.F.: The power of suggestion: inertia in 401(k) participation and savings behavior. Quart. J. Econ. **116**, 1149–1187 (2001)
21. McKenzie, C.R.M., Liersch, M.J., Finkelstein, S.R.: Recommendations implicit in policy defaults. Psychol. Sci. **17**, 414–420 (2006)
22. European Consumer Centre Network (ECC-Net): ECC-Net Air Passenger Rights Report 2011. European Consumer Centre Network, Brussels, Belgium (2011)
23. European Union: Directive 2005/29/EC of the European Parliament and of the Council of 11 May 2005 Concerning Unfair Business-To-Consumer Commercial Practices in The Internal Market. Unfair Commercial Practices Directive (2005)
24. European Parliament: Council of Common Rules for the Operation of Air Services in the Community 1008/2008 (2008)
25. Monk, A., Wright, P., Haber, J., Davenport, L.: Improving your human-computer interface a practical technique. Prentice Hall, New York (1993)
26. Holzinger, A.: Usability engineering methods for software developers. Commun. ACM **2005** (48), 71–74 (2005)
27. Nielsen, J.: Usability Engineering. Morgan Kaufmann, San Francisco (1993)
28. Cotton, D., Gresty, K.: Reflecting on the think-aloud method for evaluating e-learning. Br. J. Educ. Technol. **37**, 45–54 (2006)

Processes of Creating Infographics for Data Visualization

Mateusz Szołtysik

1 Introduction

Nowadays the popularity of infographics (Fig. 1) is increasing, it is more and more common to present content in a visual way. Even though it might seem like that, infographics are not a new idea. The concept of data visualization in the form of maps or illustrations has been known for centuries [16].

Information graphics, commonly called "infographics", are used to communicate complex data in a captivating way. In general, infographics can be described as a compilation of one or more visualizations that have been modified in order to present a specific data and highlight the significant points [3, 5].

Through infographics readers get an overview of a topic due to data and other available information. Infographics are one of the most effective means of telling stories about the data, which are captured by the reader through principles of graphic design [9].

Infographics can be a powerful visual approach to transmitting information and supporting conceptual, theoretical understanding because people "see with their brains" [18]. Due to infographics people easily remember the content because of presentation of large amount of data in a visual form [29]. It's a type of image that combines data with design, thus helping individuals and organizations communicate their message [25].

A prior version of this paper has been published in the ISD2016 Proceedings (http://aisel.aisnet. org/isd2014/proceedings2016).

M. Szołtysik (✉)
University of Economics in Katowice, Katowice, Poland
e-mail: mateusz.szoltysik@ue.katowice.pl

167

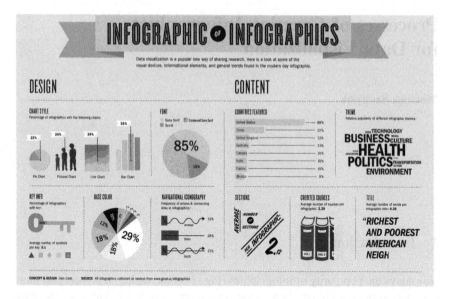

Fig. 1 The attempt to explain what an infographics are by example of one of them. *Source* [6]

The creation process and its specific elements, which can be as different as the approach of every person responsible for designing infographics might be a very interesting aspect of studies.

According to Andy Kirk, a data visualization specialist, the planning in creation process is important. He stated: "Establishing the goal of a visualization or info-graphic is to consider its developments process" [13]. Person that is responsible for design has to focus on a strategic plan, concept and all the necessary elements to create such a process.

2 Literature Review

As mentioned in the previous section, the presentation of data through graphic sym-bols is not a new idea. The history of visual presentation begins with [4, 7, 19, 25]:

1. Cave paintings, that are dated back to 15000–18000 B.C.E. In the caves at Lascaux, France it was found ca. 2000 images, which narrate stories through motion, characters etc. It is one of the oldest evidence in the world, that demonstrates power of visual data presentation.
2. Egyptian Hieroglyphs, that are dated back to ca. 3000 B.C.E. A large, picto-graphic murals responsible for communication of complex ideas to people.

3. Public Speaking, that is dated back to 500 B.C.E. The Greeks pioneered the study and practice of oratory. Centuries later the art of public speaking (Ars Oratoria) was a mark of professional competence in Rome.
4. Stained Glass Windows, that are dated back to ca. 950 C.E. The Roman Catholic Church transmitted history of saints and biblical characters to a mostly unlettered public through the colorful stained glass.
5. Bar Graphs, that are dated back to 1350 C.E. Bishop Nicole Oresme created a "Proto-Bar graph" for plotting variables in a coordinate system.
6. Comic Strips, that are dated back to 1850 C.E. Rudolphe Töpffer, a Swiss artist developed the precursor to today's modern comics. He used frames that contain both images and text to storytelling.
7. Overhead Projector, that is dated back to 1945 C.E. Police used overhead projectors for identification. It was followed by the military, educators, and businesses.
8. 35 mm Slide Presentations, that are dated back to 1950 C.E. It enabled professionals to communicate ideas sequentially to larger audiences.
9. PowerPoint, that is dated back to 1987 C.E. The new era of data presentation started with this computer program. It allows everyone to design slides and presentations.

Information graphics or infographics are graphic visual representations of information, data or knowledge. These graphics present complex information quickly and clearly [17]. The infographic is part of data visualization [26]. Data visualization includes signs, photos, maps, graphics and charts, it presents complex data [24].

Six types of visual representations that help to communicate content to the reader defined by Nancy Duarte are [7]:

1. flow (e.g., linear, circular, divergent/convergent, multidirectional);
2. structure (e.g., matrices, trees, layers);
3. cluster (e.g., overlapping, closure, enclosed, linked);
4. radiate (e.g., from a point, with a core, without a core);
5. pictorial (e.g., process, reveal, direction, location, influence);
6. display (e.g., comparison, trend, distribution).

These types of visual representations can help efficiently, precisely, and clearly transfer abstract ideas, complex and compact content that would otherwise require a long narrative. Notably when they are used together as building elements for communication. As Duarte stated, "Rather than oversimplifying the complexities… (the use of these types of visual representations) can often incorporate multiple parameters, telling a richer story of cause and effect (or any other relationship) than data points alone" [7].

The foundation and core of infographics are composed of three important parts. Those are Visual, Content and Knowledge [21]. Visual representations of data, information, and/or knowledge are presented by elements on Fig. 2:

Fig. 2 Three core parts of
infographics. *Source* [21]

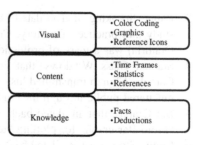

In opinion of Dave Gray and Juan Velasco infographic is a visual explanation
thanks to which the reader is able to understand, find or do something more easily; it
also integrates words and pictures in a dynamic, easy to remember way. Gray and
Velasco claim that it stands alone, is universally understandable and self-explanatory,
as well as it reveals information that was invisible or submerged [8].

Visuals are powerful tools for learning, they help to improve memory and recall.
A set of abilities that enables an individual to effectively find, interpret, evaluate,
use, and create images and visual media is referred as visual literacy [2]. They are
necessary to understand and analyze the contextual, cultural, ethical, aesthetic,
intellectual, and technical components involved in the production and use of visual
materials by the recipient [2].

The color of infographic is an important element of information provided by this
visualization method [1]. Colors influence on the audience at different levels. The
individual perception of color is derived from cultural background. It depends on
power of message assigned to the distinct color [4, 15].

But infographics can also be a threat [1]. As mentioned in earlier Huff's studies
[11] about usage of charts to inform or misinform readers, those aspects are also
actual nowadays, in the new tools of data presentation. The choice of ranges on
graphs can have huge impact on interpretation, as well as proportion of Y-axis to
X-axis also can distort the data. Huff has stated, that bar charts and pictorial graphs
should have areas proportional to values - to make comparisons only in one
dimension.

According to Edward Tufte, a statistics professor at Yale University, well known
in the area of information design and data visualization, if a graphic element of a
design in an infographic does not play a significant role to obtain specific information, it should be excluded from the final infographics design. Tufte stated "(...)
it's wrong to distort the data measures—the ink locating values of numbers—in
order to make an editorial comment or fit a decorative scheme" [27].

Tufte describes those useless design elements as "chartjunks". In this terminology he understood redundant shapes, lines or ornamental elements.

He criticized the inclusion of visual embellishment in charts and graphs as well
as considered that chartjunks deflect audience's attention and also devastate their
perception of information design [14, 27].

However Nigel Holmes had different opinion in area, where embellishment of
data graphics is encouraged. He stated that data graphic "must engage the reader's

interest" [10]. He demonstrates in his work how to achieve this by using graphic representations [3].

In his opinion "the purpose for making a chart is to clarify or make visible the facts that otherwise would lie buried in a mass of written materials" [10].

Some researchers criticize the usage of third dimensional charts. In the work of Schonlau and Peters small, but significant decrease in accuracy was discovered after addition of third dimension to the pie charts [22]. Also Siegrist stated that three dimensional bar and pie charts usually take more time to evaluate, but the accuracy of those evaluations was on the same level as two dimensional versions [23].

3   Methodology

The study was originated to discover what motivates people to design infographics and what are the components of the creative process. Those components are very important and their understanding in the future will help to unify the entire process of data visualization by example of infographics.

This study involved first and second year students that study at the Faculty of Informatics and Communication at the University of Economics in Katowice. The four groups of students were examined and every laboratory group was split into smaller teams (Fig. 3).

Fig. 3 Groups of students working at their tasks

Each study-team obtained a several sheets of paper, pencils, colorful marker pens and one from four prepared sets of statistical data, titled Topic 1, Topic 2 etc.

The data sets are referring to the year 2015 and describe issues mentioned below:

- the Silesian Police summary from 2015 [12],
- the summary concerning level of knowledge of foreign languages in Poland [30],
- the summary concerning people digitally excluded in Poland [28],
- the summary concerning level of reading books in Poland [20].

The example data set is presented on Fig. 4.

Each of the student teams has the same task: to create a draft of infographic from obtained different statistic data and to describe the entire process.

In this case they had to fill up the simple question form and answer six questions:

1. Describe the entire process which guided you to create infographic. Indicate its most important points.
2. What influenced the choice of colors used in your project. Do you consider this aspect important?
3. Do you consider that type of information presentation (i.e. infographic) is more readable for its recipients?

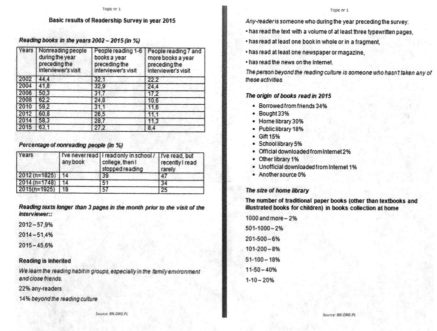

Fig. 4 One of the prepared sets of statistical data

4. Did you make drawings in your project? If yes, justify and describe the specific position of its placement in the project. If not, justify and describe why didn't you consider a drawing usage in the project.
5. What's the title of your project? If any? Do you consider this element important? Justify your answer.
6. Do you think that the subject of obtained data and the data itself have had an impact on the difficulty level of your project preparation? Justify your answer.

4 Results

This section contains the overview of obtained results. Each aspect of the study is discussed separately. Due to limitation of the study to Polish-speaking groups only, all the key parts of infographics were translated in the forms of rectangles.

4.1 Elements of the Process

In the analysis of the first aspect concerning the participants' designing processes, 7 from 12 teams have reported the most important elements of their thinking, discussion, preparation and creating process in the form of points.

One of the groups rated every point of their preparation process with the percentage values of its priority degree, giving to each the equivalent value of 25%:

• Grouping the information,
• Selecting the most important information,
• Selecting the most appropriate graphics corresponding to the given data,
• A graphical layout (simplicity).

Another team pointed three factors: transparency, readability, intuitiveness, which in their opinion are the most important in the process of infographics creation. The first impression regarding the analyzed data was also significant for that team and led them through their concept changing for obtaining more clarity and readability.

Simplicity, a short time of interaction with the data by the reader and aesthetics of the project itself were relevant factors for another group.

Other group have tried to use a simple iconography because these are clear symbols, easy to understand and associate with the facts. Pictograms were joined by statistical data with some brief descriptions. Ease of interpretation and possibility of translation infographics into foreign languages was important. That potential translation process couldn't affect the form of project and understanding of its data.

Another team identified three stages of their process:

- Analysis of the data, reading it together in the group,
- "Brainstorm" on graphic conceptions and choosing the best one collectively,
- Implementation in form of sketch by the person assigned at the beginning.

In this approach each member had to introduce its own idea, widely discussed by the other team members. The sketch has evolved to the final version of colored project design. Assignment of responsibilities was an important stage in the creating processes.

An important factor for one team was the level of topic importance: whether it is worth to discuss it. They had to choose the information to present a leitmotiv of their project in the form of a recognizable sign.

4.2 Significance of Color

An another purpose of the study to consider was the level of color significance used in infographics and its meaning for participants of study. Each team participating in the study should indicate, what influenced the choice of colors used in their project and whether they consider this issue to be important.

For most of study groups this issue has a significant position in process of infographics design. The colors are relevant, they have cultural background, derived from the evolution of human existence. For example, the green color is often associated with something positive, and red with something negative.

All participants' decision was to choose more aggressive colors to indicate negative objects. They have referred to the origins of colors as main factor in their selection, giving as an example the red color again. In their opinion, it has a negative significance because that color is widely associated with blood.

Yet another team pointed a subliminal message as an element inseparably connected with color code. That groups' opinion is that the authors of infographics are responsible for reception of their work. They can influence the audience perception of projects. With black and white drawings each person can understand them on their own way. With color we can decide on the perspective of interpretation for our audience and dictate an intentional interpretation to them. Another team chose bright colors for the titles in order to simultaneously intrigue, attract and absorb a potential recipient. In their opinion, the colors used in the infographics are relevant, because it unintentionally affects the way of receiving information by the viewer. For example, properly used color will bring up the feeling of peace and inner-balance.

Two groups have indicated that the correlation color-topic is extremely important. Both groups have used bright, eye-catching colors related to the topic of statistical data. For example, the colors that in their opinion were associated with road sign indicating frequent accidents (Fig. 5). Participants referred to attracting recipients attention about this sign and for highlight the significance and message in infographics.

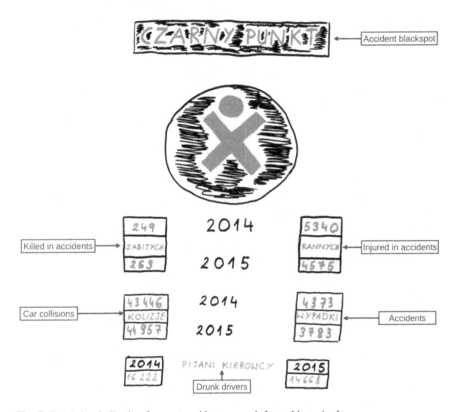

Fig. 5 Road sign indicating frequent accidents as an infographic main theme

4.3 Readability of the Message

In the next step of this study, the participants needed to determine whether that type of communication of information through infographics is in their opinion more readable for message recipients?

All the participants have agreed with the legibility of infographics as a data medium.

One group has stated: "Yes, because it is a faster and easier method to information absorption and only the most important information are extracted. That contrasts with the long tables filled with data, which overwhelm the users."

Another team has considered that: "The infographics are a lot more legible than plain text, because people process the images quicker than only-text data. That affects the level of better understanding of presented subject."

Yet another team has agreed with that statement: "By the use of many graphics, illustrations and the content comparisons contained in the infographics, they are much clearer and transparent for interpretation than in case of the text data only."

Next group has stated: "Definitely yes, as indicated by the different studies, only 20% of messages are communicated via words, and as many as 80% via the images and gestures. However, most people process the data by the visual type of learning and that method is easier to communicate the messages for them."

One group has considered, that: "Pictures and pictograms are easier to read than text, because the text data are perceived as more time-consuming. The infographics form of presentation is conducive to the rapid flow of information, however it is more suitable for the people, who are better adopted for such type of data, i.e. young people or people familiar with the Internet. Older people prefer to pay attention to the long texts, they are able to focus on what they know, in contrast with younger people."

Another team has agreed, that this way of presenting information is both more interesting and easier to assimilate. Many people may feel the desire to absorb the presented information, even despite the fact, that at the first glance data presented in the form of infographics are not from a field related with area of their interest.

Another group has considered, that original artwork attracts more attention and it allows a better understanding of the provided information, because in most cases people's perception is guided by sight and visual objects.

Members of another team have suggested, that dry facts described in a documents, either in the form of tables or descriptions, lead the recipient to get lost in what he or she is reading. In most cases it doesn't offer such benefits and effects that should be. Meanwhile, data represented by the graphics are easier to understand, assimilate and image.

Other team has decided, that visual aspect of infograpics is important and has agreed with discussed hypothesis. They referred to usage of colors, which attract the reader's attention, the design and layout of infographics itself, e.g. pie charts.

4.4 Significance of Drawings

Another purpose of the study to consider was importance of drawings in projects of infographics. The participants of study were supposed to indicate if the position of drawings in design of infographics is important or irrelevant, as well as show if the graphics itself as a data source are irrelevant.

All participants have agreed with importance of drawings in data presentation.

One group has used drawings, because of the connotation that is connected with every graphic symbol. Other team used drawings to show difference between statistical representatives of presented subject, in this case each group of knowledge of languages (Fig. 6).

Another group used drawings to reduce time of reading the data and shorten the time of infographics assimilation.

Yet another team has suggested, that usage of drawings increases the readability level and attracts attention to the communicated message, because of that they used many drawings in their project, as it can be seen on Fig. 7.

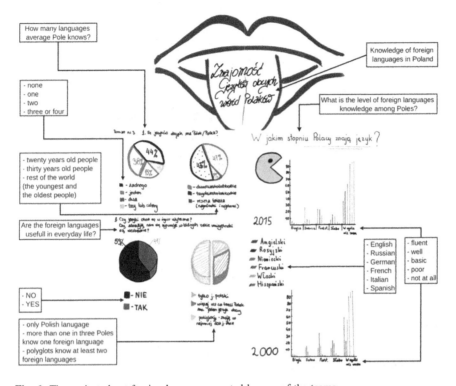

Fig. 6 The project about foreign languages created by one of the teams

They also considered that infographics with drawings are received by viewers with greater pleasure in contrast to text data. That allows to increase user engagement to the presented data.

The usage of images allows to add variety to the form of communication. Data becomes more attractive for the customers and viewers pay more attention to the data.

They used drawings to visualize almost every important statistical or percentage values and added some variety to the projects.

That approach allows to reproduce the assigned values in an interesting and comfortable way without usage of the standard data tables. Any similarities to the presented data topic were supposed to strengthen the connotation of infographics and attract the viewers. In this form the data are very structured and legible.

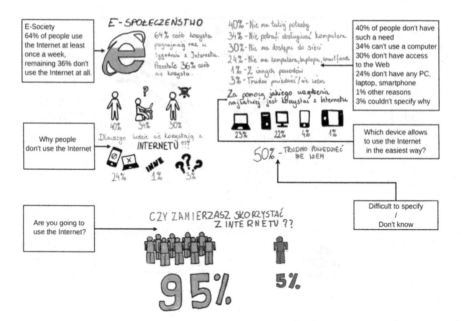

Fig. 7 Drawings in infographics attracts attention to the communicated message

4.5 Significance of the Title

In the next step of this study, participants were supposed to consider the importance of the title in infographics designs.

All participants have indicated the title as a requirement, because a person automatically interprets certain information in a certain way. Nevertheless, not all the teams (Fig. 8) have included this element in their projects.

Other teams (Fig. 9) have used title in a creative way that will catch the attention of younger readers.

Title determines the discussed issue, without title presented data could be misinterpreted. Through its use we can easily intrigue a person to view the entire project and present what a particular infographics are about. A catchy and curious title aims to encourage the viewer to read the content.

Participants have agreed that the title should reflect the data and results, it needs to be connected with the source data, to give the viewer the whole picture of presented subject. It shows what the viewer can expect from reading the following infographics.

Infographics' title is a matter of high importance, because in a few words summarizes the issues discussed later in its content, expresses the essence of the information and its subject. The title gives us a brief explanation concerning the main subject of all elements used in a infographics project. It gives meaning to the whole work, in other words: the headline summarizes the entire project.

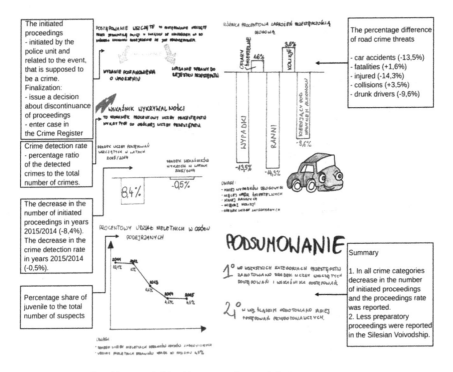

Fig. 8 Infographic without a visible title can confuse recipients

4.6   Meaning of Data

The last aspect of the study was that the subject of obtained data and the data itself have had an impact on the difficulty level of project preparation. 10 out of 11 study groups indicated that these elements are strongly correlated. Only one group completely disagreed with this statement. They have chosen to present the data in form of charts thus in their opinion it does not require a lot of effort.

Different group has agreed with the statement about the major importance of the data subject, because they are familiar with the subject of their data and thus creating the infographics was easier for them.

Another group has also agreed with this discussed aspect of study. As a justification they relied on the fact that some kinds of information does not have connotations or their connotations can be easily mistaken. The creator need to make an additional effort in the design process of creating a clear, recognizable and distinct drawing as a one part of infographics. It is essential for not misinterpreting the entire message by the viewers.

Yet another team has stated, that each data subject needs a separate artwork and a different approach to its subject.

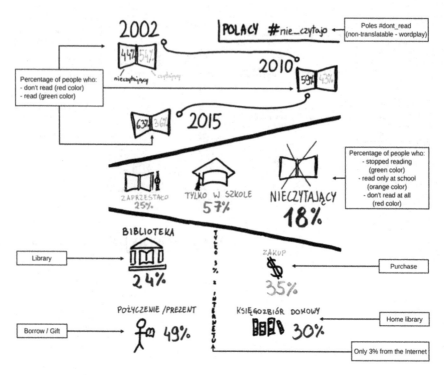

Fig. 9 Infographic title as a hash tag element

Another group (Fig. 10) has suggested, that electronic version of infographics would be simpler in creation, because in all of their statistical data were hidden additional data, i.e. in one group of respondents were included another divisions.

They gave example of mouse usage in that scenario: when user would hover the mouse cursor over the one of the statistical data, an another list of information would appear, and then it vanishes when user would move the mouse in another direction.

Other team has considered, that this is definitely matter of much significance. In their opinion the obtained topic can't be presented in a light and pleasant way. The project should emphasize the importance of the topic, its relevance and seriousness.

Yet another team has also suggested, that obtained topic gave them a wide room for maneuver. The frames of their subject were very extensive and they could develop it with many different approaches.

A significant role was marked by the level of obtained text: data written in simple and intelligible language were easier to process by participants of study. E.g. one group of participants suggested, that the type and amount of their data were too detailed, what has increased level of difficulty in their project.

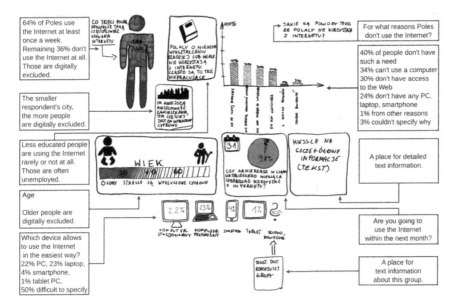

Fig. 10 Infographic design as an idea for a future electronic version

Infographics as a form of data presentation categorically affects the difficulty of preparation. It requires many additional steps in creating process. First we need to analyze all the data and then present them in interesting graphic form.

Participants, who were not earlier familiar with subject of data discussed by them, have had greater difficulty in the preparation of projects.

An another factor, which had directly influenced the difficulty level was style of composition, that had to be kept in the entire project and should correspond with the presented data.

5   Conclusion and Future Directions

After analysis we can identify the essential elements of the data transformation process from statistic data into visual data, presented by elements on Fig. 11.

Every group had pointed out that creating a good and readable infographic is a main target. We can indicate the clarity factor as significant. Selection of data, content (drawings and sketches), colors, which are responsible for the character and intensity of the process, appeared in all responses. Without earlier analysis of obtained data, they couldn't begin any operations: firstly they had to make a choice what information are worth to present.

The colors are meant to emphasize the importance of provided information. Selection of colors was often dictated by the aesthetic reasons. Colors affect the

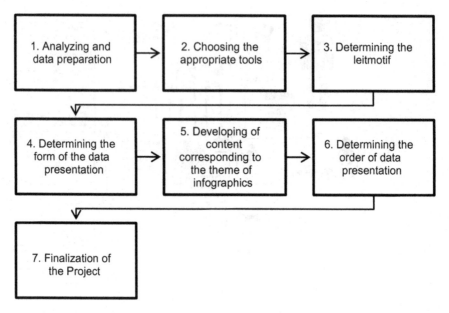

Fig. 11 The steps of participants' creating process

perception of the presented data. They affect the emotions and focus of recipient. The participating teams wanted to present the data in more readable form for the recipients. In aspect of color importance, the simplicity and readability are pointed again as significant issue in process of infographics creation. These elements are necessary for unambiguous interpretation of visual data by the audience.

The main factor that appears in all statements in favor of infographics as a medium for a readable presentation of the data is a fact, that data presented with this method can be easily memorized by the recipients.

In the era of Big Data the use of infographics as a main visualization tool to present data obtained through Big Data processes could be an another interesting direction to further research. Apart from the usage of infographics in education discussed in this paper, the adoption of informational and educational aspects of infographics as an element assisting decision support systems could also be verified. For example in the area of visual analytics for making smarter decisions faster.

An important aspect for future studies is why people didn't refer and cite the sources of obtained and presented data. Infographics as an element of data visualization must be reliable. Examining the reasons why the participants have often omitted this fundamental element in infographics, can help us creating infographics as more certain data sources. In the age of information, one of the most popular statements is: "don't believe in everything you read on The Internet".

Easy access to data resources, the availability of various types of data and the possibility for every person to be an author and creator, can be an tremendous progress for us, but also a threat. Understanding why people don't pay attention to citing the original sources will help to develop an effective approach in the education process and to stressing and increasing the level of significance of this element. Especially when everyone can share anything.

References

1. Arslan, D., Toy, E.: The visual problems of infographics. Glob. J. Humanit. Soc. Sci. **1**, 409–414 (2015). http://www.world-education-center.org/index.php/pntsbs. Accessed 20 Nov 2016
2. Association of College and Research Libraries: ACRL Visual Literacy Competency Standards for Higher Education (2011). http://www.ala.org/acrl/standards/visualliteracy. Accessed 22 April 2016
3. Bateman, S., Mandryk, R.L., Gutwin, C., Genest, A.M., McDine, D., Brooks, C.: Useful junk? The effects of visual embellishment on comprehension and memorability of charts. In: Proceedings of the ACM Conference on Human Factors in Computing Systems (CHI 2010), pp. 2573–2582. ACM, Atlanta, GA, USA (2010)
4. Biecek, P.: Odkrywaj! Ujawniaj! Objasniaj! Zbior esejow o sztuce przedstawiania danych. Fundacja Naukowa SmarterPoland.pl, Warszawa (2016)
5. Borkin, M.A., Vo, A.A., Bylinskii, Z., Isola, P., Sunkavalli, S., Oliva, A., and Pfister, H.: What makes a visualization memorable? IEEE Trans. Vis. Comput. Graph. **19**(12), 2306–2315 (2013)
6. Cash, I.: Infographic Of Infographics (2011). http://www.cashstudios.co/infographic-of-infographics. Accessed 19 Mar 2016
7. Duarte, N.: Slide: Ology: The Art and Science of Creating Great Presentations, pp. 45–58. O'Reilly Media, Sebastopol, CA, USA (2008)
8. Gray, D., Velasco, J.: What is an infographic? (2007). http://communicationnation.blogspot.com/2007/04/what-is-infographic.html. Accessed 7 April 2016
9. Harrison, L., Reinecke, K., Chang, R.: Infographic aesthetics: designing for the first impression. In: Proceedings of the 33rd Annual ACM Conference on Human Factors in Computing Systems (CHI '15), pp. 1187–1190. ACM, New York, NY, USA (2015)
10. Holmes, N.: Designer's Guide to Creating Charts and Diagrams. Watson-Guptill Publications (1984)
11. Huff, D.: How to Lie with Statistics. WW Norton & Company, New York, NY, USA (2010)
12. Informacja o stanie bezpieczenstwa i porzadku publicznego styczen - grudzien 2015 r. wojewodztwoslaskie (2016). http://slaska.policja.gov.pl/download/35/295645/stan-bezpieczenstwa-styczen-grudzien-2015.pdf. Accessed 5 Mar 2016
13. Kirk, A.: 10 Things You Can Learn From the New York Times Data Visualizations Comments (2012). http://blog.visual.ly/10-things-you-can-learn-from-the-new-york-times-data-visualizations/. Accessed 20 Nov 2016
14. Lankow, J., Ritchie, J., Crooks, R.: Infographics: The Power of Visual Storytelling. Wiley, New Jersey (2012)
15. McCandless, D.: Colours in culture. (2009). http://www.informationisbeautiful.net/visualizations/colours-in-cultures/. Accessed 20 Nov 2016
16. Marcel, F.: Infographics and data visualization tools to engage your language learners. Contact TESL Ontario **40**(1), 44–50 (2014)
17. Newsom, D., Haynes, J.: Public relations writing: form & style. Wadsworth Series in Mass Communication and Journalism, pp. 220–221. Cengage Learning, Boston, MA, USA (2010)
18. Oetting, J.: The science behind why our brains crave infographics (In an Infographic) (2015). http://blog.hubspot.com/agency/science-brains-crave-infographics. Accessed 25 Mar 2016

19. Pietrzak, M.: Zastosowanie metod kartograficznych do tworzenia infografiki edukacyjnej. In: J. Morbitzer & E. Musiał, eds. Czlowiek, media, edukacja. Uniwersytet Pedagogiczny, Krakow, pp. 272–280 (2014). http://www.ktime.up.krakow.pl/symp2014/referaty_2014_10/pietrzak.pdf. Accessed 27 Nov 2016
20. Podstawowe wyniki badan czytelnictwa za rok 2015 (2015). http://bn.org.pl/download/document/1457976203.pdf. Accessed 5 Mar 2016
21. Roy, S.: The Anatomy of an Infographic: 5 Steps to Create a Powerful Visual (2009). http://spyrestudios.com/the-anatomy-of-an-infographic-5-steps-to-create-a-powerful-visual/. Accessed 7 April 2016
22. Schonlau, M., Peters, E.: Graph comprehension: an experiment in displaying data as bar charts, pie charts and tables with and without the gratuitous 3rd dimension. RAND Working Paper WR-618 (2008)
23. Siegrist, M.: The use or misuse of three-dimensional graphs to represent lower-dimensional data. Behav. Inf. Technol. **15**(2) (1996)
24. Siricharoen, W., Siricharoen N.: How infographic should be evaluated? In: Proceedings of the 7th International Conference on Information Technology (ICIT 2015), pp. 558–564. Al Zaytoonah University of Jordan, Amman, Jordan (2015)
25. Smiciklas, M.: The Power of Infographics: Using Pictures to Communicate and Connect with Your Audiences. Que Publishing, Indianapolis (2012)
26. Thatcher, B.: An Overview of Infographics. Illinois Central College Teaching & Learning Center (2012). http://www.slideshare.net/ICCTLC2. Accessed 20 April 2016
27. Tufte, E.: The Visual Display of Quantitative Information. Graphics Press (1983)
28. Wykluczeni cyfrowo. http://www.tnsglobal.pl/archiwumraportow/files/2015/10/Wykluczeni-cyfrowo-komunikat.pdf (2015). Accessed 5 Mar 2016
29. Yavar B., Mirtaheri M.: Effective role of infographics on disaster management oriented education and training. In: Proceedings of the 27th DMISA Conference on Disaster Risk Reduction 2012, Disaster Management Institute of Southern Africa, Tzaneen, Limpopo Province, South Africa (2012)
30. Znajomosc jezykow obcych. http://www.tnsglobal.pl/wp-content/blogs.dir/9/files/2015/06/K.041_Znajomosc_jezykow_obcych_O05a-15.pdf (2015). Accessed 5 Mar 2016

Technical Consequences of the Nature of Business Processes

Václav Řepa

1 Introduction

Over 20 years after the idea of business process-driven management was born, essential insufficiencies in the understanding of how business processes should be modeled in order to respect this idea still exist. These insufficiencies influence even the existing modeling standards. As creating the model of business processes is a necessary first step in the process of implementing the idea of process-driven management, this situation should be regarded as a main obstacle to the need for putting this idea into real life.

On the other hand, there are theories, methods and ideas that are already known in other contexts, which directly address the problems and features behind the above-mentioned insufficiencies of the current state. As usual, the root of the problem is not the insufficient basis of knowledge but rather the insufficient recognition of the proper context.

The aim of this paper is to draw attention to the essential features of business processes and business systems in the context of their modeling. The needed reflection of these essential features in the modeling language and methodology will be analyzed. In addition, the basic insufficiencies of the contemporary approaches to the business process modeling will be pointed out together with the outline of possible ways to overcome them.

Methodology for Modeling and Analysis of Business Processes (MMABP) [2, 11, 14], forms the basis of the main principles used in dealing with the

A prior version of this paper has been published in the ISD2016 Proceedings (http://aisel.aisnet.org/isd2014/proceedings2016).

V. Řepa (✉)
University of Economics, Prague, Czech Republic
e-mail: repa@vse.cz

© Springer International Publishing Switzerland 2017
J. Gołuchowski et al. (eds.), *Complexity in Information Systems Development*,
Lecture Notes in Information Systems and Organisation 22,
DOI 10.1007/978-3-319-52593-8_12

considerations in this paper. MMABP is a 'language independent' methodology based on a set of meta-models which define the basic concepts and express the basic principles of the methodology, and completed with the set of techniques, consistency rules and patters. MMABP is generally open in terms of its principal ability to be complimented with newer concepts, principles, techniques, etc. if they are consistent with its principles and the meta-models. As MMABP is based on meta-models instead of particular languages, it can also be used as the base for the evaluation of any modeling language towards the principles.

The paper is divided into three main sections. After this introduction, we briefly explain the main features of the nature of business processes following the root ideas of the business process management in the second section. Based on the idea that business processes represent the integration of both managerial and technical aspects of the business, we look at it from various points of view of management as well as cybernetics. The third section contains some reflections on the consequences of the main findings from the second section. We argue there for process states and consequential rules for the process description granularity together with the integrating idea of service-oriented approach to the business system conception as a general 'technical' consequence of the idea of process driven management. In the conclusion, we then go on to briefly summarize the main ideas of the paper and outline some other consequences which call for further elaboration.

2 The Nature of Business Processes

The idea of a process-based organization is wonderfully expressed in [8]. The authors argue for so-called 'Business Process Re-engineering' (BPR) which means a complete radical change in the way organizations are managed. The proposed way of managing an organization is based on the idea that an organization has to build its behavior on an objectively valid structure of its business processes to be able to fully exploit the possibilities offered by the technology progress. This condition is typically not fulfilled in traditionally managed organizations where hierarchical organization structure prevents seeing, as well as managing, the crucial process chains which should be the central subject of change due to the technology progress. For achieving the needed ability to fully exploit the technology progress the traditional hierarchical way of management should be rejected and substituted with a management style based on the objective model of business processes of the organization.

Information technologies (IT) play the primary role in this radical revision of the traditional approach to organizational management, as IT has a double function in this process. On the one hand, the turbulent development of IT can be regarded as a determining reason for the necessary changes in organizational management in order to allow the organizations to be flexible enough to exploit the new possibilities in technology. On the other hand, IT is a primary tool for this radical change; IT allows managing organizations in ways that have been previously impossible. In essence, it allows handling businesses in a way that is closer to its natural substance, i.e. more simple than traditional ones. IT undertakes the role of the main lever for developing the business.

The essential importance of IT in BPR also works as the basis for the typically mistaken idea that BPR is a clearly technical matter. Even if this idea can be met very often, especially in the field of process modeling (unfortunately including also modeling languages, namely Business Process Model & Notation (BPMN) [1]), it is fatally dangerous in the context of BPR. Just the essential importance of the technical aspects of the business in process-oriented organizations causes any underestimation of non-technical aspects of business processes can completely destroy the effect of the process-oriented management. Therefore, we regard the technical consequences of the nature of business processes in business process modeling methodologies critically important.

2.1 Business Process from the Point of View of Management

One of the main ideas stated in the previous section is that an 'organization has to build its behavior on objectively valid structure of its business processes to be able to fully exploit the possibilities offered by the technology progress'. This idea presumes that there are some objective facts, conditions and rules, following from the business system itself, which determine the quality and correctness of business processes. The collection of such rules is can be called 'system causality'. Nevertheless, a business system is not defined only by rules but also by the behavior of its actors. System causality can restrict but not fully determine the behavior of business actors. The reason for the behavior of these business actors is not just that they can but that they need or just simply want to. Thus, in order to make the model of the business system complete, it is also necessary to take into account the concept of intentions, purposes and business goals. Not only the system causality but also *intentions* of actors should be modeled.[1]

Business system causality is the main subject of the object-oriented approaches to business processes which can be found mainly in the field of 'process ontologies', as the ontological point of view is naturally object-(system-) oriented. Some works oriented on the methodical aspects of business processes modeling can be found in [5] and [3] for instance. A special kind of these ontological attempts is

---

[1]The problem of intentionality versus causality is a serious philosophical topic which also contains the problem of 'free will' and other topics, which are still 'live' in the philosophical community. Moreover, in the context of business system modeling, this problem also covers the essential differences among various concurrent approaches to the basic issues of the theoretical economics. Although this problem would generally require more discussion, for the purpose of this paper let us reduce it just to the need of taking into the account in the business process modeling also the intentions of actors.

highly relevant for our problem, as it also takes into the account the intention and is so-called 'goal-oriented' business process modeling. In [4] the explanation of this approach can be found: '*While traditional approaches in business process modeling tend to focus on 'how' the business processes are performed (adopting a behavioral description in which business processes are described in terms of procedural aspects), in goal-oriented business process modeling the proposals strive to extend traditional business process methodologies by providing a dimension of intentionality to the business processes*'. Although this approach clearly distinguishes between the system and the process view of processes, it is still focused just on some particular aspects of processes and especially do not take into the account their collaboration.

To understand the 'business essence' of the collaboration of processes one primarily has to differentiate between two basic functional types of processes: key ones versus support ones. As customer needs are constantly changing, the processes in the organization should change as well. This means that any process in the organization should be linked to the customer needs as directly as possible. Thus, the general classification of processes in the organization distinguishes mainly between:

- *Key processes*, i.e. those processes in the organization which are linked directly to the customer, covering the whole business cycle from expression of the customer need to its satisfaction with the product/service.
- *Support processes*, which are linked to the customer indirectly—by means of key processes which they are supporting with particular products/services. To understand the 'business essence' of the collaboration of processes one primarily has to differentiate two basic functional types of processes: the key versus support ones.

Main differences between key and support business processes can be found in Table 1.

The value of the key process is given by its direct contact with the value for the customer as it is the main goal of the process. The values of other (support) processes are given by the services by which these processes support other (supported) processes. In this way every process is ultimately connected to the value for the customer either directly (key process) or through its services for other processes.

Table 1 Essential differences between key and support business processes

	Key process	Support process
Customer needs	Fulfilled directly	Fulfilled indirectly, through key processes
Responsibility	Management-oriented Responsible primarily for the context of the whole business case from the customer point of view	Production-oriented Responsible for the quality of its service, not for the context in which it is used
Dynamics	Very dynamic, often changing, permanently developing, every instance is an original	Mostly static, stable, offering standardized and multiply usable services

As it follows from the previous paragraph, the whole system of processes of the process-managed organization is a network (a nonhierarchical net structure) which is *customer-centered*. All processes mutually collaborate on the basis of support services which all are finally targeted through the key processes to the value for the customer. Traditional, principally non flexible, hierarchy of organization units is replaced with the collaborative, principally flexible, customer-centered network structure of business processes. All the system is tied together by services by which the processes mutually support each other. Therefore, such systems can be called a *'service oriented view of business'*.

2.2   Business Process from the Point of View of Cybernetics

In the previous sub-section we discussed the need for intentionality in business processes. Intentionality, or more traditionally purposefulness, is also an important topic for the ideas connected with the field of technology in such phenomena like Business Process Management Automation in general, particularly robotics and similar fields. In the legendary article [12], which is usually regarded as the root of cybernetics, the authors expressed the idea which essentially influenced the later development of cybernetics: *'all purposeful behavior may be considered to require negative feed-back'*. The concept of negative feed-back is explained there as follows: *'...the behavior of an object is controlled by the margin of error at which the object stands at a given time with reference to a relatively specific goal. The feed-back is then negative, that is, the signals from the goal are used to restrict outputs which would otherwise go beyond the goal'*.

According to the basic work in the field of process-driven management ([8]), business process always follows some goal. The goal is a fundamental attribute of a business process as it is regularly used in matured methodologies like in [7] for instance. That means that *business process is always an intentional process*. By the term 'intentional process' we mean the process of purposeful behavior of an interested object following some goal. For instance, if we personalize the business process to the behavior of its actors, namely of the process manager, we can undoubtedly see his/her behavior as an intentional behavior which follows the goal of the process.

When drawing conclusions from the previous two paragraphs, one can find that every business process, as it is an intentional kind of process, have to have some negative feed-back which ensures restriction of its outputs in order to keep them in the margins of its goal. This characteristic strongly distinguishes the business process from processes in general (i.e. in just the technical/physical sense) as well as from processes which do not need any feed-back like machine-managed or fully automated processes running without a contact with their environment.

3   Main Resulting Features of the Modeling of Business Processes

3.1   System Versus Process-Oriented Model of Business Processes

In sub-section *Business Process from the Point of View of Management* the principal need to distinguish between key and support processes is mentioned as a basic condition for understanding the 'business essence' of the collaboration of processes. This difference in process types is a system attribute, i.e. it is not visible from the process described just as an algorithm (like with use of BPMN [1]). It requires looking at the whole system of processes. Therefore, it is necessary not only to model the process as a process (i.e. how to run it) but also as a part of the system of processes which is a collection of collaborating processes mutually connected with services. MMABP calls this model the Global Process model. As a system view, this model shows the system parts (business processes) and their mutual relationships (cooperation) and that way it allows the needed functional differentiation of processes; clearly distinguishing between the key and support ones (see Fig. 1).

Unfortunately, this need is still not sufficiently reflected by the current BPM methodologies. It is quite visible also as the state of the art of business process modeling languages. For example BPMN [1], even if it is established as a worldwide standard in the field of business processes modeling, it is still mainly oriented just on the description of internal algorithmic structure of a business process and disregards the global view on the system of mutually cooperating processes. The only way of modeling the cooperation of different processes in BPMN is using 'swimming pools and lanes' in the Collaboration Diagram.

Unfortunately, the global aspects of the system of business processes cannot be sufficiently described this way nor its completeness ensured. The BPMN primarily views processes as sequences of actions in the time line. However, the global model requires seeing processes primarily as objects (relatively independent of the time), distinguishing different kinds of them (especially the key vs. support ones), describing their global attributes (like the goal, reason, type of customer, etc.), and recognizing their essential relationships to other processes which all is obviously impossible to describe as a process flow.

One of the mostly accepted 'de facto' standards which fully support the system (object-oriented) view of business processes is the Eriksson-Penker Notation [7]. It was created as an extension of Unified Modelling Language (UML) [18] which corresponds with the 'object nature' of the global view on processes discussed above. This notation distinguishes between the 'Business Process View' which illustrates the interaction between different processes and the 'Business Behavioral View' which describes the individual behavior of the actors of one particular process. This way it respects the important difference between the global object-oriented view of a process system and the detailed process-oriented view of a single process. Therefore, the MMABP methodology presented in this paper uses

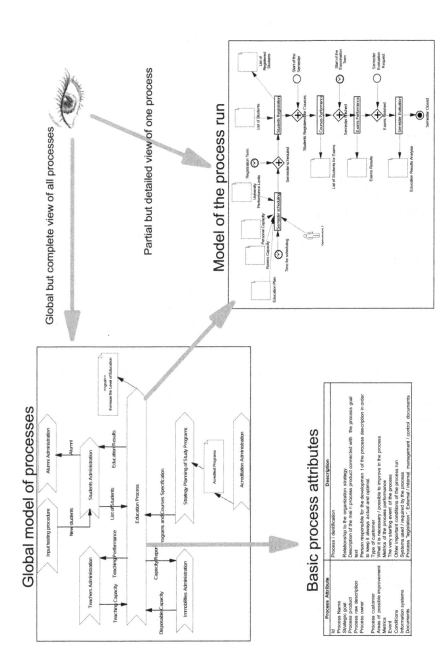

Fig. 1 System process view versus a detailed view of process. *Source* MMABP

the Eriksson-Penker process diagram as a complement to the BPMN in order to compensate for the absence of the global view in this language (see Fig. 1). A detailed explanation of the methodical need for global model of processes as well as related criticism of the BPMN can be also found in [13].

3.2 Process States

One of the basic topics discussed in section *The Nature of Business Processes* and viewed in both its sub-sections from managerial and technical points of view is *the intentionality in business system*. Sub-section *Business Process from the Point of View of Cybernetics* then concludes that to be a business process the process must be connected with its environment via so-called 'negative feedback'. The 'negative feedback' is implemented in MMABP as the concept of a *process state*.

In the case of the business process the feed-back is represented by the input to the process from its environment which is causally connected with some process output. The value of the input should influence the following behavior of the process in terms of keeping it within the margins of its goal. This means that 'intermediate' inputs to the process (i.e. none-starting inputs to the process coming between its starting and end points) are critically important parts of the business process distinguishing it from other, non-intentional (i.e. non-business), processes. When working with processes we have to take into the account even the time dimension; every input to the process from its environment has to be synchronized with the process run. Thus, in each part of the process where some input which influences the following process run is expected the process state has to be placed. The process state means such points in the process structure where nothing can be done before the input to the process occurs, i.e. the point of waiting for the input.

The *Process state* thus represents the essential need to synchronize the process run with expected events. This need follows from the fact that the event is always an objective external influence and thus it must be respected. From the physical point of view such respect means synchronization—waiting for the event. As BPMN do not recognize the concept of a process state there is no other way than to express the process state with the general symbol for synchronization—the 'AND gate'. In order to distinguish between the general synchronization and its specific meaning as a process state, we complete the BPMN with the stereotype ≪process state≫.

One of the most important ideas following from the concept of a process state is that there cannot be a sequence of process steps uninterrupted by the process state. This rule reflects the essence of the definition of an *elementary process step*:

(a) the process step is regarded as elementary if there is no objective reason for its interruption,
(b) the reason for the interruption of the step is objective if it comes from outside of the process.

Rule (b) of this definition means that each objective reason for the process interruption is represented by an event (external influence). Thus, any step of the process, no matter how technically complex it is, must be regarded as elementary if there does not exist an external influence (event) which the process has to respect (i.e. wait for). This consequence clearly illustrates the fact that the elementariness of a business process step is not only its physical but much more a functional attribute as the business process itself is always more than a physical process (algorithm) only. This way the methodology prevents the analyzer from the pointless unlimited dividing of the process activities which is a frequent mistake in the field of business process modeling. The necessity of such a safety fuse in the methodology against the unlimited division of activities is given by the fact that in the field of process-oriented modeling the aggregation is a dominating type of abstraction (unlike in the field of object-oriented modeling where the generalization is a dominating type of abstraction). This fact manifests itself in the principally unlimited possibilities of division of activities known as a rule: any single process activity can be decomposed into the structure of sub-activities—a process. As the division of activities is physically unlimited, the methodology has to define some logical—functional definition of the very low level: the level of the process elementariness.

Figure 2 shows an example of the process with states represented with the symbol of synchronization (AND gate). It also illustrates the rule of 'needed objective reason for the process interruption'. For instance, the first two states in this process represent waiting for the action of a customer. The objectiveness of this state is obvious, as the customer can never be regarded as a part of the enterprise. From the enterprise point of view, the customer is always an independent actor, a representative of free will. Therefore, the process cannot continue unless the will of the customer is known [either via a direct action or as information gathered indirectly by means of the timer (see events in process states in Fig. 2)]. The third state represents waiting for the action of another—supporting—process (service). Even in this situation the reason for waiting is objective although anticipated events come from inside of the enterprise. In this case the objectiveness is given by the transfer of the responsibility. For the service the supporting process is responsible, not the main one. Thus, this state represents the collaboration of processes (see also the sub-section *Collaboration of Processes*).

3.3   *Granularity of Process Description*

The MMABP rule of 'needed objective reason for the process interruption' discussed in the previous sub-section also works as a determiner of the 'proper' granularity of the process description. In fact, it addresses just one from the four process abstraction levels which MMABP distinguishes:

194   V. Řepa

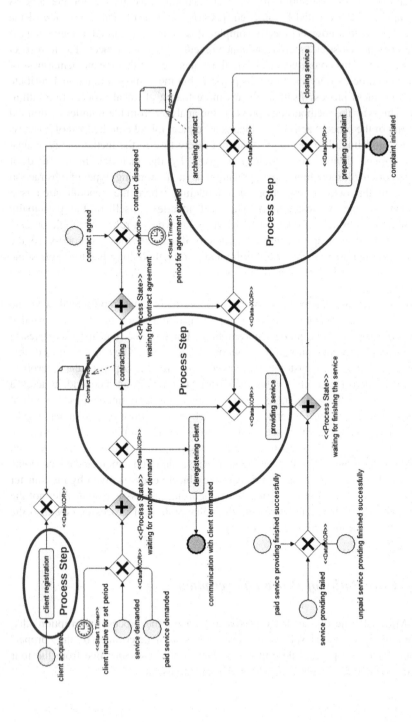

Fig. 2 Example of the use of process states in the BPMN language. *Source* author

1. *Enterprise functionality level* (functional division of the organization to different process areas according to different key processes of the organization).
2. *Business process level* (Process Map—Global model of processes).
3. *Process step level* (process description with process states according to the rule of 'needed objective reason for the process interruption').
4. *Activity level* (more detailed decomposition of process activities according to the states of crucial business objects).

A detailed description and explanation of four process abstraction levels can be found in [17]. From the characteristics of particular process levels it is evident that activities, mentioned generally in the previous sub-section correspond to the so-called 'process steps'. Figure 2 illustrates the process model on the level of process steps. The process step can consist either of one activity (see the *client registration* step) or even more time-independent activities connected to one structure (see both remaining process steps). In this example, both structures represent mutually exclusive activities. Exclusivity ensures their time-independence as at the given moment only one activity exists. Another possible structure, which is not illustrated there, is a structure of parallel activities (connected with the AND gate). Even parallel activities have to be regarded as time-independent, because they all run at the same time. The only incorrect structure is a sequence of activities which is a representative of the clear time-dependency (i.e. the time of one activity is dependent on the time of another one).

Regarding the reasons for as well as crucial circumstances of using process states it is obvious that the *'process step'* (as well as the *process state*) primarily reflects *collaboration*. This topic is discussed in detail in the following sub-section.

3.4   Collaboration of Processes

Figure 3 shows different problem areas connected with the process-based organization. Three exemplary viewpoints at the figure together address three substantial parts of the organization's life: content, technology, and people. Each particular point of view is characterized by typical questions which should be answered by the methodology in the given field.

From previous sub-sections follows that the common concept visible in all discussed topics is the collaboration of business processes. The need for having two basic types of models: a system and detailed, process-oriented ones is caused by the need to distinguish between two basic functional types of processes: key and support ones. The reason for distinguishing between key and support processes arises from the need for creating such structures of mutually collaborating processes which ensure that all processes follow the same goal: value for the customer. Every support process is connected with value for the customer through supporting the other processes. Therefore, the most important aspect of the detailed model of the process (i.e. model of the process run) is the identifying those points in the process

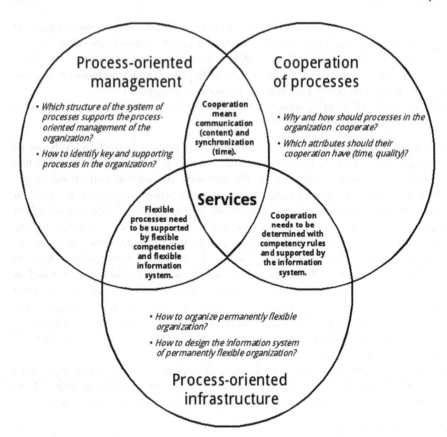

Fig. 3 Service as a common denominator of content, technical, and human aspects of the organization management. *Source* [15]

which represent the collaboration with other processes—identifying process states. Finally, *process states play the role of the determiner of such granularity of the process description which is driven mainly by the collaboration with other processes.*

Following the idea of the service-oriented view of business expressed in the sub-section *Business Process from the Point of View of Management,* we can see that the common denominator of all important aspects of the organization management is the service (see Fig. 3). The concept of service connects both the managerial and technical points of view of the business process. It plays a crucial role in understanding of the functional meaning of processes (key vs. support processes) as well as in understanding their mutual positions (collaborative network instead of the control hierarchy). The concept of service also plays the important role from the technical point of view on business processes:

- The MMABP business process modeling technique uses the service as a universal form of the description of mutual relationships between two collaborating processes. Every pair of relationships between two collaborating processes is described as a single service in terms of *request for the service* and *service delivery*. Subsequently, the details of the service are specified using the standard attributes of the service taken from the general theory of Service Level Agreement (SLA). That allows the central expression of all important aspects of the organization management connected with the collaboration of two particular processes in once: content of the service together with the human aspects in the form of mutual responsibilities of actors, and also the technical aspects like technical parameters of the service (time and data requirements, etc.).
- In the process of implementing the system of processes, the service plays the role of the precise definition of technical as well as human and other organizational requirements which should be handled and supported with the technology.

4   Conclusions

This paper is a reflection on the nature of business processes and the way which their nature necessarily manifests itself in business process modeling methodologies and languages. It points out the most important aspects and principles of business processes management and tries to reflect them in the features of the technical and methodical support of the business process modeling. We can summarize these main features as follows:

- The necessity to model the process system as well as process details.
- Two basic types of models: the global (system-oriented) and the detailed (process-oriented, algorithmic) should be created separately but in mutual connection.
- The necessity to model the states of the process and to recognize different, mutually completing events.

The cybernetic principle of 'negative feed-back' causes recognizing the process states. The necessary negativeness of the feedback then requires modeling mutually completing events (i.e. waiting for more than just one event) in order to reflect the fact that there always must be some objective reason for the decision about the further run of the process (in order to restrict the further process run to the relevant direction in terms of the process goal).

- The necessity to keep four levels of the granularity of process description (abstraction levels).
- The necessity to pay the proper attention to the collaboration of processes.

The outputs and states of two collaborating processes should be taken together as parts of one service (as a superior concept covering both) which allows the permanent integration of all important aspects of the process management in the process model.

The paper also critically evaluates existing business process modeling standards, mainly BPMN. It is because BPMN can be regarded as a worldwide de facto official standard and also because it contains, unfortunately, most insufficiencies. Other popular and significant standards (Eriksson-Penker [7], ARIS [16], IDEF [10]) are not so problematic like BPMN. Nevertheless, they still cannot be regarded as completely perfect from the point of view of outlined necessary features. For instance, although process states are present in all of them some way, ARIS does not clearly distinguish between them and events, Eriksson/Penker methodology as well as IDEF take them just as a technical issue.

The topics discussed in this paper also represent serious challenges for the future development of business process modeling and management methodologies, with a further elaboration on the topic of intentionality in business processes seeming to be the most important. This topic represents the direct relationship between the managerial and technical meaning of business process management and thus it is critically important for its meaningfulness. The paper already mentions the important influence of the ontology engineering area by this topic ([4]). In other connected areas, especially in Philosophy, the concept of intentionality is elaborated in more detail which also actually calls for implementing it in the field of business process modeling languages ([6, 9]).

Acknowledgements The paper was processed with support of long term institutional support of research activities by Faculty of Informatics and Statistics, University of Economics, Prague.

References

1. Business Process Model and Notation (BPMN). OMG Document Number: formal/2011-01-03, Standard document URL: http://www.omg.org/spec/BPMN/2.0 (2011)
2. Business System Modeling Specification: http://opensoul.panrepa.org/meta-model.html
3. Cabral, L., Norton, B., Domingue, J.: The business process modelling ontology. In: 4th International Workshop on Semantic Business Process Management (SBPM 2009), Workshop at ESWC 2009, 1 June 2009
4. Cardoso, E.C.S., Almeida, J.P.A., Guizzardi, G., Guizzardi, R.S.S.: Eliciting goals for business process models with non-functional requirements catalogues. In: Proceedings of 10th International Workshop on Business Process Modeling, Development and Support, CAISE 2009, vol. 29, pp. 33–45, Amsterdam, The Netherlands (2009)
5. De Nicola, A., Lezoche, M., Missikoff, M.: An ontological approach to business process modeling. In: Proceedings of the 3th Indian International Conference on Artificial Intelligence, 17–19 Dec 2007
6. Dennett, D.C.: The Intentional Stance, MIT Press, (1988). ISBN: 0-262-04093-X
7. Eriksson, H.E., Penker, M.: Business modeling with UML: business patterns at work, Wiley (2000). ISBN: 0471295515

8. Hammer, M., Champy, J.: Re-engineering the corporation: a manifesto for business revolution. Nicholas Brealey Publishing, London (1993)
9. Heyes, C.M.: Contrasting approaches to the legitimation of intentional language within comparative psychology. Behaviorism **15**(1), 41–50 (Spring 1987)
10. Mayer, R.J., Menzel, C.P., Painter, M.K., deWitte, P.S., Blinn, T., Perakath, B.: IDEF3 process description capture method report, knowledge based systems, Inc. (1997)
11. OpenSoul project: http://opensoul.panrepa.org
12. Rosenblueth, A., Wiener, N., Bigelow, J.: Behaviour, purpose and teleology. Philos. Sci. **10**, 18–24 (1943)
13. Řepa, V.: Business process modeling notation from the methodical perspective. In: Towards a Service-Based Internet, pp. 160—171. Springer, LNCS, Berlin (2011). ISBN 978-3-642-22759-2
14. Řepa, V.: Business system modeling specification. In: Chu, H., Ferrer, J., Nguyen, T., Yu, Y. (eds.). Computer, Communication and Control Technologies (CCCT '03). Orlando: IIIS, pp. 222–227 (2003). ISBN 980-6560-05-1
15. Řepa, V.: Cooperation of business processes – a central point of the content, technical, and human aspects of organization management. In: Proceedings of the BIR 2013 Conference, Springer, LNBIP, Warsaw, pp. 78–90 (2013). ISBN: 978-3-642-40822-9
16. Scheer, A.W: Architecture of integrated information systems: principles of enterprise modeling. Berlin et al. (1992)
17. Svatoš, O., Řepa, V.: Working with process abstraction levels. In: Proceedings of the BIR 2016 Conference, Springer, LNBIP, Prague, pp. 65–79 (2016). ISBN: 978-3-319-45320-0
18. Unified Modelling Language™ (OMG, UML) Infrastructure Specification, version 2.4.1, Object Management Group (OMG) (http://www.omg.org)

The Goals Approach: Agile Enterprise Driven Software Development

Pedro Valente, Thiago Silva, Marco Winckler and Nuno Nunes

1 Introduction

Software development within enterprises still lacks accuracy, and effectiveness is still far from being achieved as project full-success rates are still as low as 30% [1], and there is still a long bridge to cross until software development within enterprises is achieved in a patterned way, and established as a consistent source of revenue following investment within enterprises [2]. Nevertheless, the advances of Software Engineering (SE) have at least taken us from a chaotic state of the practice [3], to a more inspiring situation where enhanced executive management support, agile methods, and increased user involvement are appointed as factors for software project success [4].

Our work is inspired by the need to improve software project success rates within enterprises, where the establishment of a tool that enhances communication capabilities between both Enterprise Engineering (EE) and Software Engineering

A prior version of this paper has been published in the ISD2016 Proceedings (http://aisel.aisnet.org/isd2014/proceedings2016).

P. Valente (✉)
University of Madeira, Funchal, Portugal
e-mail: pvalente@uma.pt

T. Silva · M. Winckler
Université Paul Sabatier, Toulouse, France
e-mail: rocha@irit.fr

M. Winckler
e-mail: winckler@irit.fr

N. Nunes
Madeira-ITI, Técnico — U. Lisboa, Lisbon, Portugal
e-mail: njn@uma.pt

© Springer International Publishing Switzerland 2017
J. Gołuchowski et al. (eds.), *Complexity in Information Systems Development*,
Lecture Notes in Information Systems and Organisation 22,
DOI 10.1007/978-3-319-52593-8_13

(SE) knowledge-based expertise can be seen as crucial for the effectiveness of the Software Development Process (SDP) that may be applied. However, this enhancement can only be achieved if a common framework of shared concepts of the business and software domains is established and used to build the Information System, which today can be seen an inherent part of the global enterprise system.

We present the *Goals Approach*, which focuses on tailored in-house development of Information Systems for Small and Medium Enterprises, which is characterized by needs of agility concerning the supportive SDP in order to allow the achievement of tangible results in limited amounts of time and budget [5]. *Goals* defines a SDP that applies a straightforward method that analyses the enterprise in a top-down process in order to elaborate a business model, called as Enterprise Structure. And continues by detailing the Enterprise Structure components using cross-consistent concepts in order to design the User Interface, the Business Logic and the Database, including the Enterprise Structure as the back-bone of a final Software Architecture, which can be used for in-house software development management.

Briefly, the *Goals* conceptual structure (back-bone Enterprise Structure components are underlined) includes: the human interaction which is represented by means of Business Processes, User Tasks, User Intentions and User Interactions; the User Interface which is represented by Aggregation Spaces, Interaction Components, Interaction Objects, and Interaction Spaces; which (the last one) can also be used by its Business Logic, which is composed by Business Rules, User Interface and Database System Responsibilities; and the Database which is composed by Data Entities and Fields.

This paper focuses on the validation of the cross-consistency of concepts that supports each component, and provides insight on how each can be implemented. The related work to our approach is presented in Sect. 2, the *Goals Approach* SDP and Structure are presented in Sect. 3. The method is presented in Sects. 4 (Analysis Phase) and 5 (Design Phase), the cross-consistency validation is presented in Sect. 6, and the conclusions and future work are presented in Sects. 7 and 8.

2   Related Work

Considering enterprise-driven development in the EE domain, our approach is distinct from the DEMO-based GSDP [6] as it provides a structured user interface specification. e3Value [7], is a method which also relies on the GSDP, and models the business for value adding the "value interface" inputs and outputs, yet, also not providing a user interface elaboration solution. Still in EE, our approach can be compared to Archimate [8] and BPMN [9] in the perspective that it provides an enterprise and software structuring language. It is however different in the perspective that it applies a method to specify a business model and derive a software architecture.

Our approach can be compared to the business-oriented 'Management by projects' [10], ITIL [11], and the SE' SCRUM [12] and XP [13] methods, which also define techniques and architecture for BPI, yet, none of these methods specifies the software architectural pattern that should be used. *Goals* can be used by these methods for software architectural specification, and in the cases of SCRUM and XP for the specification of architectural spikes regarding iterative implementation, matching agile software architecture [14], as *Goals* further structures the enterprise business model.

Regarding the Human-Computer Interaction perspective, the closest solutions are methods that settle for user interface conception based on user task and domain models [15, 16]. Our approach is different as it complementarily conceives the Business Logic layer based on enterprise business regulations and coordination structures.

3   Software Development Process

Goals is an Enterprise-Driven Human-Centered Software Engineering (HCSE) method that bridges enterprise requirements and software implementation by means of a business model. It introduces the Interaction Space as the space that supports both in-person and remote interaction whilst applying the same business regulations and data concepts. The Interaction Space bridges the Business Processes and its User Task's human interaction, and establishes a relation with the Business Rule and Data Entity concepts in order to architect the Enterprise Structure (the business model) which is the back-bone of Software Architecture.

Goals was developed for over a decade of applying the HCSE *Wisdom* Approach [17] in software practice in a SME. *Wisdom* provides the (original) definition of Interaction Space (IS) and the architectural technique that bridges human interaction and system behavior (based on the IS), expressed by User Tasks and Business Rules in the Enterprise Structure, and by User Interactions and System Responsibilities in the Software Architecture. It uses DEMO [18], for the definition of Business Process and Business Rule, *Activity Modeling* (AM) [19], for the cornerstone definition of User Task. And uses *Hydra* [20], for the User Interface definition of Aggregation Space, and BDD [21] for system behavior specification, both concerning the Software Architecture elaboration.

The *Goals Approach* Software Development Process (SDP) integrates the Enterprise Engineering (EE) and Human-Computer Interaction (HCI) perspectives in the process of defining a Software Architecture for a given Business Process Improvement (BPI) in two phases: the Analysis Phase which elaborates the Enterprise Structure, and the Design Phase which elaborates the Software Architecture.

The Analysis Phase identifies Business Processes (BP) in Step 1, User Tasks (UT), in Step 2, Interactions Spaces (IS) in Step 3, Business Rules (BR) in Step 4, and Data Entities (DE) in Step 5, composing the Enterprise Structure. The Design

Phase details and complements the Enterprise Structure by means of a User-Centered Design (UCD) perspective, specifying each UT by means of a Task Model (Step 6), designs the User Interface (Step 7), and structures the Business Logic (Step 8) and the Database (Step 9), finishing with the elaboration of a final Software Architecture (Step 10), given an MVC architectural pattern [22].

The process continues with the Implementation and Testing Phases (which detail is out of the scope of the present paper), and uses the Software Architecture to guide the software development, and the User Interface Design, Task Model and User Stories to guide the Information System test before deployment. Figure 1 illustrates the SDP Analysis and Design Phases using a BPMN diagram [9], and each EE, HCI and SE domain's contribution and cooperation suggestions for each Step.

The Enterprise Structure is the *Goals* business model, and is composed of Business Processes (BP) and its User Tasks (UT), which are Essential Use Cases [19]. Actors communicate by means of Interaction Spaces (IS) when carrying on their UTs, which apply Business Rules (BR) that represent business regulations which are applied over used Data Entities (DE). Each component is identified in a top-down methodological process, and its definition, origin and symbol is presented in Table 1.

Briefly, the Software Architecture is composed by one Aggregation Space [20] per UT, which is composed by Interaction Components and Interaction Objects that trigger User Interface and Database System Responsibilities (SR), which architecturally use the ISs, BRs and DEs associated to that UT ensuring traceability between business and software. Each Software-Specific component is presented in Table 2.

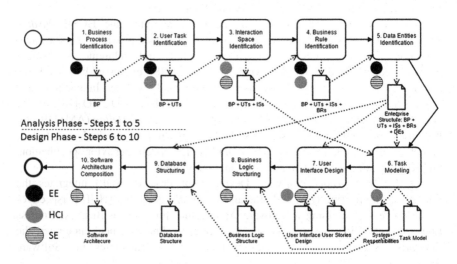

Fig. 1 *Goals* software development process

Table 1 Enterprise Structure components definition, origin and symbol

Component	Definition	Origin	Symbol
Business Process (BP)	A set of UTs that lead to a Goal	DEMO	
User Task (UT)	A complete task within a BP	AM	
Interaction Space (IS)	The space that supports a UT	Wisdom	
Business Rule (BR)	A restriction over DE's structural relations	DEMO	
Data Entity (DE)	Persistent information about a business concept	Wisdom	

Table 2 Software architecture software-specific components

Component	Definition	Origin	Symbol
Aggregation Space (AS)	A User Interface	Hydra	
Interaction Component (IS)	Tool of a User Interface	Goals	
Interaction Object (IO)	A User Interface Object that triggers SRs	Goals	
User Interface SR (UI SR)	A SR that provides support for User Interface presentation	Goals	
Database SR (DB SR)	A SR that manages Data Entities	Goals	

4 Analysis Phase

The Analysis Phase defines a top-down methodological process that identifies and relates each Enterprise Structure component in five Steps, which are presented in Sects. 4.1 (Step 1—Business Process Identification), 4.2 (Step 2—User Task Identification); 4.3 (Step 3—Interaction Space Identification); 4.4 (Step 4—Business Rule Identification); and 4.5 (Step 5—Data Entity Identification).

4.1   Step 1—Business Process Identification

Goals defines a Business Process (BP) as "*A set of User Tasks that lead to a Goal*". The Goal is the objective, and also names the BP. It is expressed as a unique set of related enterprise business concepts (Data Entities) which support the BP execution, and that will compose the enterprise domain model as will be presented in Step 5— Data Entity Identification. The relation between the BP and the set of managed business concepts increases awareness on the problem begin solved, and also the communication capability between project stakeholders by means of the in-depth of their knowledge on the specific part of the enterprise that is being evolved. This facilitates the BPI development, and in practical terms results in faster and more productive project meetings, increasing the probability developing projects in fewer time.

The relation between BPs and Data Entities is useful to design the enterprise BP Model, which relates BPs, Actors and Data Entities, increasing the perception on how a BP uses and produces certain business concepts from a higher level, which is useful for business management. We present the BP Model, by means of the application of the Process Use Cases Model [23] adapted to the current *Goals* notation. The meta-model and an example are presented in Fig. 2.

Figure 2 presents the meta-model of the BP Model, in which it can be read that only one actor can "Initiate" a BP, but an unlimited number of Actors can participate in it, and also, that an unlimited number of Data Entities can be used and produced by a BP. It also presents an example where Actor A initiates the BP, Actors B and C participate in it, and the Data Entity A is used and the Data Entity B is produced.

4.2   Step 2—User Task Identification

The User Task (UT) definition is derived from the concept of Essential Use Case (EUC) [19], which defines a Use Case as a "complete and meaningful task (carried out in relationship with a system)". This definition is adapted to the enterprise context based on the principle that the Business Process (BP) is a sequence of UTs, and that each UT is carried out by a single Actor. Since a BP always has a limited

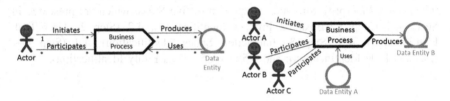

Fig. 2 Business process model meta-model, and BP model example

number of tasks, all UTs can be considered as meaningful, thus, we abandon the "meaningful" term and define a UT as "*A Complete User Task within a BP*". We also apply the principle that an Actor (a user) never carries on two UTs consecutively and separately, which is an axiom that aims user performance and software development efficiency by inducing the reduction of the articulatory distance of the UT i.e. the user's effort [24], and by suggesting that the necessary tools should be provided using as little user interface implementation space as possible. If two UTs are consecutive, then they can be merged in a single sequence of acts, expressed by a single UT, leading to is completion in the same way.

The relations between UTs are what designs a BP. The consecutive relation is the most common, as it supports the most common BP flow. Yet, it is not sufficient to represent more complex services that must be available in different interaction points (identified as touchpoints by the Service Design domain) which usually have back-end support, and may be visited by the customer, but not necessarily and always in a pre-defined order. This need for flexibility can be attained by the definition of conditional relation, and thus, we further define it (the conditional relation), meaning that the execution of a specific UT or BP path is conditioned to the will of the Actor. This reflects the case when an enterprise suggests its customers the execution of a given action in sequence of any other interaction but will never be sure that they will follow the suggestion, and yet continues to provide the remaining service.

Figure 3 presents the meta-model of the UT, in which further defines that one Actor can carry on many UTs and that a UT can also be carried out by many Actors defining cooperative collaboration; one BP can have many UTs; one UT can belong to many BPs; and UTs are related consecutively or conditionally. The example shows the initial UT being triggered by Actor A and consecutive B and C UTs being carried out by Actors B and C, and the response tasks, D and E (which path is conditional) being carried out by Actors B and A respectively.

4.3 Step 3—Interaction Space Identification

The Interaction Space (IS) definition is derived from *Wisdom* original concept of Interaction Space, as a user interface space where the "user interacts with functions, containers and information in order to carry on a task". We adapt this concept to the

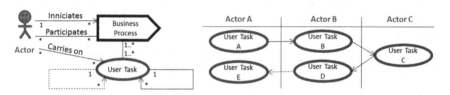

Fig. 3 User Task meta-model and example

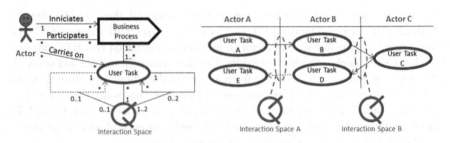

Fig. 4 Interaction space meta-model and example

enterprise context by means of its generalization, in order to complementarily consider the support of the UT in person, as in any of the cases (remote or in person), the same Business Rules (BR) and Data Entities (DE) also apply. We (re) define the IS extension as *"The Space that supports a UT (with the same BRs and DEs)"*. Hence, one IS supports the interaction between two users in person or remotely while each one carries on his own UT. Even if many UTs are carried by out many Actors in a cooperative way, the UTs will still be different. If two Actors carry on the same UT remotely, then they are performing cooperative work [25].

The identification of ISs is derived from the interaction between sequenced UTs, in order to support one Actor request and other Actor response, as in any case the same BRs and DEs apply. Figure 4 presents the meta-model that specifies that an IS supports many UTs based on the interaction between Actors, with at least a consecutive relation and at most one conditional relation.

The example shows the derivation of ISs in order to support the interaction between Actors A and B, and Actors B and C, by means of ISs A (a Request IS) and B (a Coordination IS) respectively, which is possible since the set of UTs A, B, D and E are subject to the same BRs and DEs, and the same happens in the case of UTs B, C and D. If another interaction between Actors A and B would occur (e.g. between User Task E and F), then a new IS should be defined (e.g. C) in order to support that interaction.

4.4 Step 4—Business Rule Identification

The Business Rule (BR) definition is provided by DEMO notion of Action Rule, which defines a structure of decision (using pseudo-code) that applies restrictions to Object Classes concerning the execution of business Transactions. These restrictions are paradigmatic relations (considering a semiotic association) which are applied to the syntactic relations (also considering a semiotic definition) which exist between Data Entities (DE), in order to produce a new valuable and more complex business concept. Hence, we define BR as *"A Restriction over DE's Structural Relations"*. BRs represent regulations or explicitly defined requirements that should

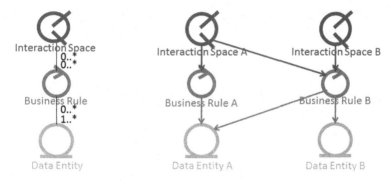

Interaction Space
0..*
0..*

Business Rule
0..*
1..*

Data Entity

Interaction Space A

Business Rule A

Data Entity A

Interaction Space B

Business Rule B

Data Entity B

Fig. 5 Business rules meta-model and example

be elicited in order to understand the restrictions which the user is subject to when carrying on a UT, and do not represent collaboration impositions with other Actors, since these rules are already expressed by the BP design.

The BRs are the grounding foundation of the Information System' Business Logic, as they are the only business-specific programmed class concerning this layer, the *middleware* of the system. The Business Logic will also be complemented in Step 8—Business Logic Structuring, with programmed parts responsible for presentation and data management.

Figure 5 presents the meta-model, which defines that an IS can use many BRs, and that a BR can be used by many ISs, and also defines that a BR can use one to many DEs, and that a DE can be used by many BRs. The example shows that IS A uses BRs A and B, and that IS B is used only by BR B. It also defines that BR A uses DE A, and that BR B uses DEs A and B.

4.5 Step 5—Data Entity Identification

The Data Entity (DE) definition is provided by *Wisdom* as a class of "*Persistent Information about a Business Concept*". This means that persistency will be maintained by the Information System, and that it will enclose meaningful concepts which are recognized within the enterprise by those who have knowledge about it. DEs are related between each other, allowing a simple representation of reality which is made available by means of a Database application. Those "meanings" enclose attributes. In terms of common database objects, DEs are expressed as tables, and attributes are expressed by fields [26]. DEs are related between each other by means of semiotic syntactic relations, which are expressed in *Goals* using a UML association [27], also implying the definition of multiplicity between the related DEs. Multiplicity will typically be of one-to-many or many-to-many. The definition of a specific multiplicity (e.g. 1–5) is uncommon, and should be expressed by a BR due its volatility (as it will eventually change). The definition of

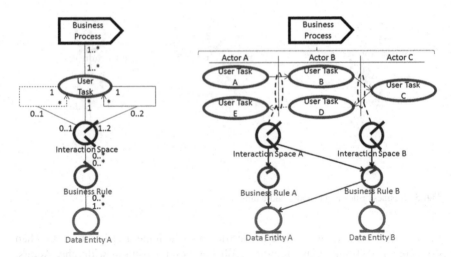

Fig. 6 Enterprise structure meta-model and example

relations of one-to-one is also uncommon as in those cases the DEs meanings can usually be conciliated in a single DE.

As mentioned in Step 1—Business Process Identification, the identification of DEs should be carried along the BP design and consequential Steps, so that the analyst develops a well-defined notion of the concepts involved in the BPI under analysis. In the current Step, the DEs only need to be identified and related to the BRs in order to compose the Enterprise Structure, the final artefact of the Analysis Phase, as illustrated in Fig. 6, with the DEs as the support of the Enterprise Structure.

The Enterprise Structure presented in Fig. 6 is composed by every identified component until this moment and also by their relation to other components, with no changes. It represents a relation which is representative of the enterprise in terms of a logic that relates Business Processes (BP), User Tasks (UT), Interaction Spaces (IS), Business Rules (BR) and Data Entities (DE) in terms of dependency and functional specification. It can be used in order to identify the implications of changing the enterprise in terms of its impact in the software structure, since, changing BPs, UTs or BRs, which is common in the business management domain, will inevitably change the underlying information system to which the 3 lower levels layers (IS, BR and DE) are an inherent part, as they are also part of the Software Architecture.

5 Design Phase

The Design Phase details and complements the Enterprise Structure with new software-specific components that build-up the Software Architecture in a top-down methodological process in five Steps, which are presented in Sects. 5.1 (Step 6—

Task Model), 5.2 (Step 7—User Interface Design), 5.3 (Step 8—Business Logic Structuring), 5.4 (Step 9—Database Structuring), and 5.5 (Step 10—Software Architecture Composition).

5.1 Step 6—Task Model

The Task Model details User Tasks (UT) in order to obtain information to carry on the User Interface design, which happens in Step 7—User Interface Design. The Task Model follows the technique applied in the *Wisdom* method in order to specify the UT in terms of User Intentions (steps that the user takes to complete the task) and System Responsibilities (that provide the necessary information), following a traditional decomposition of an Essential Use Case (EUC) [19].

The decomposition of the UT in terms of User Intentions is carried out by means of the Concur Task Trees (CTT) technique [28]. CTT defines the User Intentions in the perspective of what the user wishes to do in order to obtain what the wants from the system and complete his UT. Each User Intention has an associated System Responsibility (SR) that provides the necessary information to an Interactive Component that supports user interaction. The SR is a programmed class which is part of the Information System' Business Logic.

The Task Model is represented using an Unified Modeling Language (UML) Activity Diagram [27], defining the flow of User Intentions that lead to the accomplishment of the UT. Each User Intention uses an Interaction Component that in its turn uses a SR. These are User Interface SRs. The last User Intentions always leads to SRs that manage information, which are Database SRs. In the case when new Data Entities are identified by means of the Task Model elaboration, then they must also be represented in the DE's structure, a design task that will be specified in Step 9—Database Structuring.

Figure 7 presents the meta-model of the Task Model, where it can be read that a UT has many User Intentions, which have up to n initial User Interactions, and up to m last User Interactions that use $m + n$ Interaction Components (which compose the Aggregation Space that will support the UT). Each Interaction Component

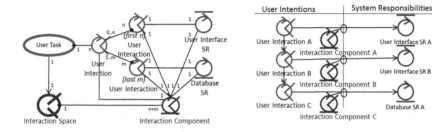

Fig. 7 Task model's meta-model and example

supports one User Intention, and uses one User Interface SR or one Database SR. The example shows the decomposition of UT A of the designed BP, which has two initial User Intentions (A and B) and one final (C). User Intentions A and B relate to User Interface SRs A and B, and User Intention C relates to Database SR A, meaning that the UT can be carried out by means of 3 interactions, which are supported by 3 System Responsibilities and 3 Interaction Components.

5.2 Step 7—User Interface Design

The User Interface Design is carried out by means of the application of the Behavior Driven Development (BDD) method [21] that further specifies each User Intention, and also frames it in terms of used Aggregation Spaces (AS), specifying the navigation between User Tasks (UT). BDD is an agile software development method that describes the system behavior based on a User-Centered Design (UCD) perspective, producing pseudo-code for User Interface specification. BDD specifies User Stories for a system feature (a UT) which is used within a certain scenario (the Aggregation Space), resulting in specific behavior which is generated by the system in the User Interface. The pseudo-code has the following syntax.

```
Given [State] When [Interaction] Then [System Behavior]
```

Where [State] represents the actual the state of the system (which identifies the Aggregation Space where the UT occurs); [Interaction] is a flow of User Interactions; and, [System Behavior] is the expected outcome that triggers User Interface and Database System Responsibilities by means of Interaction Objects. BDD also specifies the Data Entities (DE) Fields used in each User Interaction. This specification facilitates the mapping between Systems Responsibilities and DEs that occurs in Step 8—Business Logic Structuring, and the completion of the Database specification that happens in Step 9—Database Structuring. BDD's User Stories are represented by an UML Activity Diagram, and use the pseudo-code which is presented in Table 3.

Figure 8 presents the User Interaction meta-model and an example of a User Story that specifies each Task Model' User Intentions in terms of Interaction Objects that match the already identified Interaction Components (IC) and SRs

Table 3 Relation between BDD 'pseudo code syntax and software architecture' components

BDD pseudo-code	Goals component
Feature 'Feature'	User Task 'Feature'
Scenario 'Scenario'	User Intention 'Scenario'
Click, Choose, Set	User Intentions 'Click' or 'Set'
Display 'Page' or Go to 'Page'	User Interface SR 'Display Page' + AS 'Page'
Field	Data Entity Field
[Then]	(Last) System Responsibilities

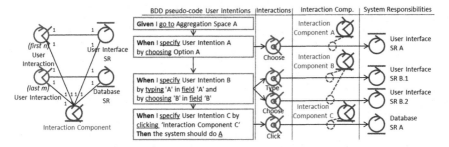

Fig. 8 User interaction meta-model and example

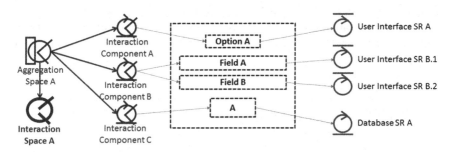

Fig. 9 User interface design example

when there is only one User Interaction. And that divide in distinct Interaction Objects when there is more than one User Interaction, as in the case of User Interface SRs B.1 and B.2 that support two Interaction Objects (for "Type" and "Choose") of IC B.

Figure 9 shows a representation of the User Interface which defines that the Aggregation Space A uses ICs A, B and C, which trigger the User Interface SRs A, B.1 and B.2, and Database System Responsibility A by means of the Interaction Objects presented in the User Interface. The relation with Interaction Space A is inherited from the Enterprise Structure.

5.3 Step 8—Business Logic Structuring

The Business Logic Structuring is carried out by defining the relations that each System Responsibility (SR) and Business Rules (BR) has to Data Entities (DE), since the relation with the User Interface components is already established at this stage.

Figure 10 shows the manual mapping that was done between SRs and DEs. BR A is inherited from the Enterprise Architecture, as also is its relation with DE A.

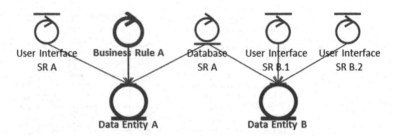

Fig. 10 Business logic structure example

User Interface SR A has been mapped to DE A, and it is assumed that Field A and B identified in Step 7, belong to DE B, which is the reason why User Interface SRs B.1 and B.2 are related to DE B. By means of the analysis of the semantic of the Database SR A, it is assumed that there was a decision to relate it to both DEs A and B.

5.4 Step 9—Database Structuring

The Database Structuring is now possible since all Data Entities (DE) are identified. Two DEs (A and B) have been identified, and DE B provides information for Fields A and B. We assume for purposes of example that DE A can only related to a single record in DE B, yet, on the contrary, any record in DE B can be related to many records in DE A. Figure 11 presents the Database Structure.

5.5 Step 10—Software Architecture Composition

The composition of the Software Architecture is carried out by relating in a single diagram every component identified by means of the execution of Steps 1–9, including the Business Process and User Tasks, and the hybrid Enterprise Structure and Software Architecture components of: Interaction Space, Business Rule and Data Entity, as well as the Software-Specific components.

Figure 12 presents the specified Software Architecture, in which the User Task (UT) A is now supported by Aggregation Space A and the underlying software

Fig. 11 Database structure example

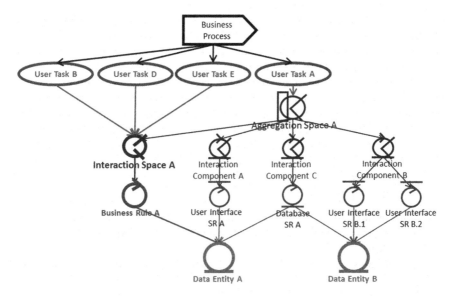

Fig. 12 Software architecture example

structure, whilst UTs B, D and E are still not automated, reason why they are directly related to Interaction Space (IS) A. UT C and IS B and Business Rule B are not represented.

The Software Architecture can be used in order to specify implementation responsibilities for a software development team and implementation priority. Priority will usually be from bottom-to-up, since the upper objects use the bottom ones. Applying the technique to the example architecture, the precedence of implementation would be: DE B (since it will be used in), DE A, Business Rule A, Database SR A, User Interface SRs B.1 and B.2, and only then User Interface SR A. Interaction Components A, B and C can follow any order, and once Interaction Space A and Business Rule A are developed, the Aggregation Space A can be implemented and tested.

6   Research Method and Validation

The research method of our approach was based on the question of if it would be possible to establish a relation between enterprise valuable concepts and the implementation of a supporting system. And by placing the hypothesis that it is possible if a cross-consistent definition of concepts is established between the business concepts that specify human interaction, and from them, derive the components of the architecture of a software system respecting specified business regulations. The cross-consistency between concepts is formalized by means of the application of the Cross-Consistency Assessment (CCA) [29] method to the Software Architecture components. Complementarily, we also use the CCA relation of concepts for purposes of architectural specification aiming software development clarification by means of providing implementation options insight.

The Software Architecture includes the five defined Software-Specific components (as previously presented in Table 2): Aggregation Space (AS), Interaction Component (IC), Interaction Object (IO), User Interface System Responsibility (UISR) and Database System Responsibility (DBSR); and the three hybrid Enterprise Structure components (which were also previously presented in Table 1): Interaction Space (IS); Business Rule (BR) and Data Entity (DE). Concerning software development, each component assumes distinct implementation options, as follows:

- **Aggregation Space** (AS). A User Interface, a Web Page that includes other Web Pages (Interaction Components), including an HTML presentation template [20].
- **Interaction Component** (IC). User Interaction for presentation and interaction support. A web Page, including an HTML template and a configuration artefact.
- **Interaction Object** (IO). A User Interface object that allows interaction. An HTML element e.g. Text Field; Checkbox; Radio Button: Dropdown List; Button.
- **User Interface SR** (UI SR). Programmed routine that supplies a recordset to be used in one or more ICs. An SQL Server Stored Procedure, View, or JAVA programmed class.
- **Database SR** (DB SR). Programmed routine that receives a recordset and saves it in the Database. An SQL Server Stored Procedure or JAVA programmed class.
- **Interaction Space** (IS). Programmed routine that can be invoked by any Software-Specific component in order to validate the data received in the User Interface, and sent to the Database.
- **Business Rule** (BR). Programmed routine that provides validation about the data which is transferred between the Interface and the Database.
- **Data Entity** (DE). Tables and Fields [26].

The Software Architecture components are presented in Fig. 13 from top to down (from the AS to DE), according to the nature of their relation of usage i.e. the component on top uses and depends on the component on the lines below to properly work [27]. We define four types of relations concerning Software Architecture specification:

- **Architectural Usage**—Underlined correct sign ($\checkmark$). *Goals* architectural relations. Define relations between components which are generated by means of the application of the *Goals* method, and which are part of its meta-model, as presented throughout Sects. 4 and 5.

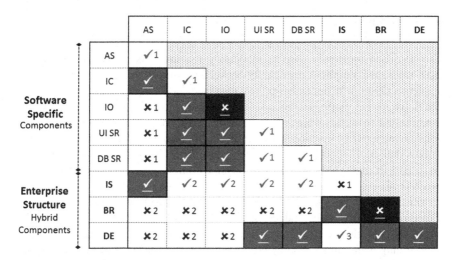

		AS	IC	IO	UI SR	DB SR	IS	BR	DE
Software Specific Components	AS	✓1							
	IC	✓	✓1						
	IO	✗1	✓	✗					
	UI SR	✗1	✓	✓	✓1				
	DB SR	✗1	✓	✓	✓1	✓1			
Enterprise Structure Hybrid Components	IS	✓	✓2	✓2	✓2	✓2	✗1		
	BR	✗2	✗2	✗2	✗2	✗2	✓	✗	
	DE	✗2	✗2	✗2	✓	✓	✓3	✓	✓

Fig. 13 CCA validation of cross-consistency

- **Allowed Usage**—Correct sign (✓). Relations that can be applied for the purpose of architectural optimization. Mostly represent: reuse (✓1), of the components by itself by means of architectural observation, IS invocation (✓2), for purpose of the data validation, or direct usage of DE by IS (✓3), meaning that no restrictions are applied in this case.
- **Contingency Usage**—Wrong sign (✗). Which are relations that that should not occur, but that yet can represent a useful trade-off, as they can simplify implementation, however introducing architectural disorganization: between Software-Specific components (✗1), or related to the Enterprise Structure components (✗2), as BRs should always be accessed by ISs and not directly, and also as DEs should be access by means of UI or DB SRs, and not by User Interface components.
- **Restricted Usage**—Wrong underlined sign (✗). Relations that should not exist, as the IO should never make use of itself, and well as the BR, in order to promote business regulations organization.

Hence, the cross-consistency of concepts between the Enterprise Structure and the Software Architecture components is achieved by means of the relation between the AS and the IS, providing support for the human interaction and business regulation execution, supporting our hypothesis. The AS also establishes the relation with the Software-Specific components by means of the Interaction Components and their Interaction Objects, namely with the User Interface and Database SRs, which use Data Entity elements, which are also common to the Enterprise Structure, providing full traceability between business and software implementation.

7 Conclusions

Our approach inherently aims at facilitating requirements elicitation, focuses on user needs, and simplifies traceability between business requirements and software implementation, which matches project management needs and user involvement in the Software Development Process, in what we believe that is the more important contribution of our work. The base strategy, based on Business Process Improvement (BPI), fits most successfully sized projects, as based on The Standish Group statistical reports, projects under 1 M$ (one million dollars), in which cost most BPI fit into, are believed to be up to 10 times more successful than 10 M$ projects [4].

Our approach is suitable for in-house development in Small and Medium Enterprises (SME), as it produces a controllable set of elements for a single BP organizational change, which will usually be implemented with great efficiency (concerning man-hours work) by programmers with knowledge of the domain, and also defines an agile and straightforward logic, which suits SME needs for development performance. This induces iterative enterprise and information system continuous development which is compatible with the Agile Development manifesto [30], at an enterprise scale.

8 Future Work

Future work mostly concerns the development of a toll for the application of the *Goals* method, as we believe from the long term use of the presented concepts, that the *Goals* structure is sufficiently well-defined in order to define a Platform-Specific Model (PIM) that can be used for full-stack MVC code generation. In this way, we open the space to predict software development effort with higher accuracy, and ultimately identify successful enterprise and software development patterns.

References

1. The Standish Group. Chaos Report 2014. (2014)
2. Valente, P., Aveiro, D., Nunes, N.: Improving software design decisions towards enhanced return of investment. In: Proceedings of ICEIS 2015, pp. 388–394 (2015)
3. Morgenshtern, O., Raz, T., Dvir, D.: Factors affecting duration and effort estimation errors in software development projects. IST **49**, 827–837 (2007)
4. The Standish Group. Chaos Report 2013 (2013)
5. Gerogiannis, V., Kakarontzas, G., Anthopoulos, L., Bibi, S., Stamelos, I.; The SPRINT-SMEs approach for software process improvement in small-medium sized software development enterprises. In: Proceedings of ARCHIMEDES III (2013)
6. Kervel, S., Dietz, J., Hintzen, J., Meeuwen, T., Zijlstra, B.: Enterprise ontology driven software engineering. In: Proceedings of ICsoft 2012 (2012)

7. Pombinho, J.: Value-oriented enterprise transformation—design and engineering of value networks. Ph.D. Thesis, University of Lisbon—IST (2014)
8. The Open Group: ArchiMate 2.0 understanding the basics. https://www2.opengroup.org/ogsys/catalog/W130 (2013). Accessed 17 Feb 2017
9. Völzer, H.: An overview of BPMN 2.0 and its potential use. In: BPMN 2010 Lecture Notes in Business Information Processing, vol. 67, pp. 14–15, Springer (2010)
10. Gareis, R.: 'Management by projects': the management approach for the future. Int. J. Project Manag. 7(4), 243–249 (1989)
11. Cannon, D.: ITIL service strategy. ISBN: 978-0113313044 (2011)
12. Schwaber, K.: Agile project management with scrum (developer best practices). ISBN: 978-0735619937 (2004)
13. Beck, K.: Embracing change with extreme programming. Computer 32(10), 70–77 (1999)
14. Grundy, J.: Foreword by John Grundy: architecture vs agile: competition or cooperation? In: Agile Software Architecture. ISBN: 978-0124077720 (2013)
15. Sousa, K., Mendonça, H., Vanderdonckt, J., Rogier, E., Vandermeulen, J.: User interface derivation from business processes. In: Proceedings of SAC 2008, pp. 553–560 (2008)
16. Sukaviriya, N., Sinha, V., Ramachandra, T., Mani, S., Stolze, M.: User-centered design and business process modeling: cross road in rapid prototyping tools. In: Proceedings of INTERACT 2007, pp. 165–178. LNCS (2007)
17. Nunes, N.: Object modeling for user-centered development and user interface design: the wisdom approach. Ph.D. Thesis, Universidade da Madeira (2001)
18. Dietz, J.: Enterprise Ontology—Theory and Methodology. Springer, Berlin Heidelberg. ISBN: 978-3540331490 (2006)
19. Constantine, L.: Human Activity Modeling—Toward a Pragmatic Integration of Activity Theory and Usage-Centered Design. Springer (2009)
20. Costa, D., Nóbrega, L., Nunes, N.: An MDA approach for generating web interfaces with UML ConcurTaskTrees and canonical abstract prototypes. In: LNCS, vol. 4385 (2007)
21. Chelimsky, D., Astels, D., Helmkamp, B., North, D., Dennis, Z., Hellesoy, A.: The Rspec Book. ISBN: 1934356379 (2010)
22. Zukowski, J.: The model-view-controller architecture. In: John Zukowski's Definitive Guide to Swing for Java 2. ISBN: 978-1430252511 (1999)
23. Valente, P.: Goals Software Construction Process: Goal-Oriented Software Development. VDM Verlag Dr. Müller, ISBN: 978-3639212426 (2009)
24. Winckler, M., Freitas, C., Palanque, P., Cava, R., Barboni, E.: Usability aspects of the in-side-in approach for ancillary search tasks on the web. In: IFIP TC13 Conference on Human-Computer Interaction 2015 (INTERACT), pp. 207–226 (2015)
25. Grudin, J.: Computer-supported cooperative work: history and focus. Computer 27, 19–26 (1994)
26. Awang, M., Labadu, N.: Transforming object oriented data model to relational data model. New Comput. Archit. Appl. 2(3), 402–409 (2012)
27. Booch, G., Jacobson, I., Rumbaugh, J.: The Unified Modeling Language Users Guide. Addison-Wesley (1998)
28. Paternò, F.: Model-Based Design and Evaluation of Interactive Applications. Springer, London (1999). ISBN: 978-1-4471-0445-2
29. Ritchey, T.: Principles of cross-consistency assessment in general morphological model-ling. In: Acta Morphologica Generalis, vol. 4 (2015)
30. Agile Alliance: Agile Manifesto. Retrieved 18 Oct 2016: http://agilemanifesto.org/iso/en/principles.html

The Main Factors Affecting E-voting Service Implementation: The Case of Palestine

Fouad J.F. Shat and Pamela Abbott

1 Introduction

Recent years have witnessed increasing debates about the implementation of e-government, and these discussions often include critical evaluations of various factors that contribute to e-government's success or failure. E-government uses information and communication technology (ICT) to facilitate interactions between the government and individuals served by the government. As ICT refers to any communication that requires technology, such as television, cell phones, and computer networks, and thus, is an umbrella term for technology-based communication services, e-government is just one form of ICT. It can be argued that the success of e-government implementation requires a consideration not only of technological change, but also of the various shifts in social, political, and cultural contexts, and the wide range of actors and institutions involved in development efforts [1]. Any research on the implementation of e-government in developing countries will need to consider a variety of fields such as organizational, socio-political, cultural, etc. and their relevance to the application of e-government. In other words, e-government is not only about technology issues but involves several political, economic, social and organisational issues [1–3].

A prior version of this paper has been published in the ISD2016 Proceedings (http://aisel.aisnet.org/isd2014/proceedings2016).

F.J.F. Shat (✉) · P. Abbott
Department of Computer Science,
Brunel University London, London, UK
e-mail: Fouadjf.Shat@Brunel.ac.uk

P. Abbott
e-mail: p.y.abbott@sheffield.ac.uk

The implementation of e-government in developing countries can be more complex than in developed countries. Therefore, different strategies may be called for that take into consideration cultural factors as well as technological ones [4–6]. In China, e-government services are still not widely accepted, and various challenges obstruct their use [7]. Scientists maintain that when advancing ICT in the developing world, it is necessary to address each of the fundamental components, rather than present an elaborate, grand-scale approach [7]. Researchers of e-government confirm that in developing countries in Asia, Africa, Europe and Latin America they are multiple problems regarding the development and implementation of e-government [8]. These problems include structural problems and system implementation issues such as determining who will manage the system, aversion to using the services on the part of the constituents, the reluctance of public employees to implement new technologies, lack of trust by all parties, cultural differences and the digital divide between generations impacting acceptance of and know how relating to technology. Researchers in the area of e-government claim that there are certain aspects that should be investigated more thoroughly such as the barriers and restrictions citizens have in accessing e-government, hindering their full participation, and how different strategies adopted by government agencies can facilitate this participation [8].

2   The Failure of E-government Projects

Stories about the success of e-government implementation abound; however, the truth is that most of the e-government projects in developing countries fail [5]. What constitutes failure can vary significantly according to context, time, and viewpoint, but e-government failure can be defined as the inability of such a system to achieve predefined goals or other, previously unanticipated benefits [5].

In developing countries, there is a large gap in the physical, cultural, economic, and various other contexts between the software designers and the place in which the system is being implemented, and this leads to incompatibilities between the system that is already in place and that which is being introduced [9].

It is clear that even though e-government brings many economic and social benefits, real challenges can hinder its establishment. Some of these challenges come from technological perspectives, and others come from social perspectives [10–12]. One example of these challenges is that e-government is expensive to implement, it requires highly skilled technicians and a solid technical infrastructure [13]. Scholars of e-government suggest the following risk factors that should also be taken into account when implementing an e-government system: political stability, an adequate legal framework, trust in government, importance of government identity, the economic structure, the government structure (centralised or not), levels of maturity within the government and citizen demand [14, 15].

The difference that exists between less-developed countries and developed countries, regarding e-government development, is highlighted in the literature. The factors that create and maintain this difference have been discussed in many studies, and a great deal of attention is paid to it by many researchers [4, 16]. Sometimes the failure of e-government implementation in developing countries has to do with the fact that guidelines based on experiments in developed countries have been followed. To alleviate this situation, a multi-disciplinary investigation should be conducted to attain a better understanding of these challenges. However, before planning, developing, and implementing e-government initiatives, it is necessary to recognise and understand the challenges that are faced by developing countries in their attempts to implement e-government projects.

3 E-voting Service Implementation

E-voting, as a, means for allowing electronic voting in a democracy, is just one example of e-government platforms. Thus, contextual issues affecting the implementation of an e-voting system will also be part of the issues affected e-government more broadly as e-voting is one component of e-government and environmental barriers and enablers of e-government implementation [8], will also likely be present in e-voting. According to the UK government, *"E-Government is not an end in itself. It is at the heart of the drive to modernise government and meet the needs of an increasingly electronic-based society. Modernising local government is about enhancing the quality of local services and effectiveness of local democracy"*. The UN model of e-government implementations includes e-voting as service of e-government for enhancing democracy and increase the direct participation of citizens and accountability [17]. Researchers describe e-voting as e-government specialty [18]. Amongst the various approaches and models that contribute to the theory and practice of e-democracy, one widely discussed component is e-voting. This is defined as any voting that involves the use of electronic means.

4 Background Research

The first motivation of this research is to provide an in-depth qualitative analysis of the main issues relating to the Palestinian implementation of e-voting, through investigating the factors which have affected the implementation of e-voting. The Palestinian political context is unique and stands to democratically advance from an e-voting system that would allow millions of Palestinian diaspora to participate in the democratic process through e-voting. The UN model of e-government implementations includes e-voting as a service of e-government for enhancing democracy and increasing the direct participation of citizens and accountability. This is particularly true in the case of Palestine, but to date, there have not been any in-depth qualitative studies that seek to understand the

Palestinian contextual environment and perceptions among the Palestinian elite in regards to factors of implementation. The interactive, democratic potential of the technology is yet to be fully utilized in the case of Palestine; the true potential for e-democracy lies in the exploitation of the internet for interactivity between government and citizens [19–21].

The second motivation of this research is to use the case of Palestine to advance the current state of knowledge on the impact of complex political environments. Despite a growing interest in the application of Information and Communication Technology (ICT) to re-structure democracy, there is a lack of research that studies the different challenges that are faced when trying to implement e-government in complex political environments [8, 22]. Using the case of Palestine, this research seeks to advance the state of knowledge on perceptions of the role of the complex political environments on implementation possibilities and outcomes. To understand the contextual environment of implementing e-democracy, one must consider the complex political situation in the democracy potentially implementing such a system and the more systems that can be examined, the greater our understanding of contextual impact will be.

In fulfilling the first and second motivations of this research, it is responding to charges to narrow the knowledge gaps on contextual factors of e-voting implementation cited in the literature [23–25]. As this research seeks to both advance the state of knowledge on the contextual factors at play in e-voting implementation in Palestinian, as well as advance the broader state of literature that addresses the role of complex political environments as barriers and enablers in e-voting, it is filling a void in the extant literature. The qualitative approach employed herein is needed to advance the state of knowledge on perceptions of e-voting. There is a line of argument stating that the empirical studies conducted on online deliberation in an e-democracy context do not provide enough grounds from which to draw solid conclusions [24]. This emphasizes the need for qualitative studies to provide greater contextual understanding. There is a need to look beyond what is found on the Internet through more quantitative content analysis or more qualitative discourse analysis and ethnographic approaches so that the democratic potential of the internet can be fully grasped [24]. This points to the need for further research to advance the theory and practice of e-democracy.

4.1 The Background of the Case Study

Since the events of 1948 termed as "nakba" (catastrophe) in Palestinian history, person self-identified as "Palestinian" have been scattered to different countries around the world. This forced diaspora creates particular challenges as the Palestinian Authority (PA) continues to seek international recognition as an emerging state. With the potential of democratic elections and a fully recognized Palestinian state in the horizon, it becomes imperative that officials within the PA find new and innovative ways to engage a scattered constituency in the democratic process [26]. The route to e-Democracy, using internet based technologies to foster citizen participation in government, could afford a displaced constituency greater access to the workings of the emerging state [27, 28].

Due to the spread of Palestinians in different countries, Palestine has experienced great difficulties in enabling its citizens to elect their representatives. Using a democratic way to elect people representatives will be an ideal to solving the Palestinian leadership crisis [26, 29]. According to government officials, many countries who are hosting Palestinian as refugees do not allow them to hold democratic activities in their territories; hence e-voting systems could offer a solution to overcome this problem [30, 31].

In June 2011, A Palestinian association called the Facilitation Office (FO) of the civic registration reported that they had settled a safe electronic voter enrolment machine for the Palestinian National Council (PNC) elections. In their statements, it was announced

> The procedures applied reflect international best practice and standards, and have been developed with the relevant national and international institutions to ensure equality of principle and practice across our exile and refugee communities, mindful of the obstacles and challenges faced by each [26, 29].

Several months later, the central election commission in Palestine declared that an e-voting system is not appropriate for Palestine. They announced that many people will not trust e-democracy [22].

5 Methodology

Based on a review of the extant literature relating to the phenomenon under consideration, the question addressed here is "What are the contextual issues affecting the implementation of an e-voting service in Palestine?" As previously demonstrated, the academics have empirically observed the need to assess this question, and they noted its paucity in the extant literature. This paper has used a grounded theory methodology (GTM) to determine the factors, herein contextual issues, that impact the success, likelihood, and outcome of the implementation of e-voting services in Palestine, Authors have used qualitative research methodology for the purpose of obtaining rich interpretive insights for the above-mentioned phenomena [21, 32–35].

To solicit the qualitative data needed to address this research question, the authors conducted 19 in-depth one-on-one semi-structured interviews with individuals identified as stakeholders in the implementation of e-voting in Palestine. Interviewees included senior engineers, state political leaders, and private sector representatives responsible or aware of the development and implementation of e-government projects in Palestine. Interviewees were selected based on positions of leadership (i.e., the level of policy influence, administrative level, or private sector leader) and were invited to participate by phone and/or email. Of the thirty-one individual from whom an interview was requested, nineteen (61.2%) agreed to participate. Interviews occurred between October 2015 and December 2015 and lasted an average of one hour, but ranged from forty-five to seventy-five minutes in length. All interviews were conducted in Arabic by the authors. The semi-structured interview instrument was developed to meet the needs of the research question, as well as adapted to be individualized to the role of the interviewee. Moreover, the instrument was semi-structured as to allow concerns and perceptions to emerge based on the issues perceived as important to the interviewee, so as not to bias the data. The interviews solicited information of the technical, procedural, and political components of implementing e-government in Palestine. For analysis purposes, the interviews were audio-recorded, transcribed verbatim, and then translated into English using qualitative research back-translation reliability methods.

The interview data produced through the 19 interviews were then analyzed using Nvivo10. The analysis of interviews presented within this paper seeks to address the research question through analyzing the data to provide meaning to perceptions of the contextual issues affecting the implementation of e-voting in Palestine. To this effect, Nvivo10, qualitative data analysis software, was used to analyze the interview data. The remainder of the section is outlined as follows: (1) introduction to the methods used to analyze the interviews, (2) provision of Nvivo outputs and analysis of the themes developed and presented in the outputs, and (3) conceptualizations of the emergent themes. Moreover, the results are supported by examples of interview responses to demonstrate how the emergent themes and concepts are grounded in the data.

Following grounded theory methods (GTM), this analysis aligns with the first two stages of coding: open coding and selective coding [34, 36]. In the open coding stage, the nineteen interviews were constantly compared and annotated to develop themes, sub-themes, and core categories. In this stage, each interview was first analyzed separately, and the results were compared across the interviews to assess similarities and differences using the constant comparative method. Memos were used in the coding process to elaborate on the data at a conceptual level, providing context to the emergent themes, to begin to develop an understanding of the relationships among the themes. Following the open coding, selective coding was used to further develop these themes into dense core categories.

5.1 GTM Open Coding

In the open coding process, each line, sentence, and paragraph of each of the transcripts was examined to label data that relates to the interviewee's perception of the possibility of the implementation of e-voting service in Palestine. These labels were applied to identify the general trends in the data and to after completion, generate a general theory of the trends. Table 1 provides the themes that emerged from the data in the first column. After all, nineteen transcripts were coded, the

Table 1 Codebook for Nvivo analysis

Theme	Meaning
Capability	Parent code for capabilities relating to the implementation and execution of e-government
Political factors	Parent code for political factors relating to the implementation and execution of e-voting
Occupation	Comments relating to the occupation of Palestine and its impact on e-voting
Corruption	Comments relating to potential or actual corruption in the government relating to implementation
Authority	Comments relating to the loss of political authority through implementation or the authority held over the project
Political will	Comments relating to the desire of the people or government to implement e-voting
Trust	Parent code for trust relating to the implementation and execution of e-voting
Trust technology	Comments that address trust of e-voting technologies and election outcomes
Trust tech people	Comments that address the human roles in e-voting technologies and election outcomes
Acceptance	Parent code for acceptance relating to the implementation and execution of e-voting

second "meaning" column was added to document the purpose of the code and its meaning to the researchers following the coding process. This codebook was thus a living document and adjusted as themes emerged in alignment with the open coding stage of GTM. Analysis memos (commonly referred to as annotations) were used to align the themes with the data being applied to them. Within the codebook, yellow themes are used to provide context (i.e., background on the interviewee that would improve the researchers' ability to understand the context in which the interviewees perceptions have developed), blue themes are parent nodes (i.e., broad/major themes), purple themes are child nodes (i.e., sub- themes to the parent nodes), which were not aggregated to parent nodes, and green themes are general themes (i.e., general concepts that did not apply to other categories for was still relevant for understanding the interviewees' perceptions of the possibility of e-voting in Palestine). Almost all data were found to be relevant to these categories with a few exceptions of data excluded for irrelevance to the phenomenon under study.

5.2 GTM Selective Coding

In the second step of grounded theory methodology, selective coding is used to identify and present the core concepts needed to understand the relationship between the themes as presented in Fig. 1. Specifically, selective coding was used to further develop these themes into dense core categories. The concept map in Fig. 1 has been obtained to demonstrate the selective coding themes. These dense

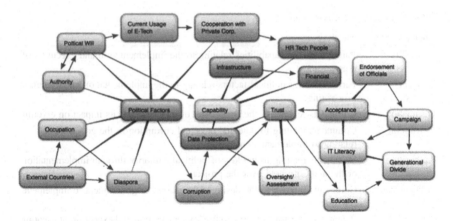

Fig. 1 Concept map of open code themes (NVIVO)

core themes, which will be used to organize the remainder of the interview findings, are as follows: Political factor, Capability, Trust, and Acceptance. The associated sub-themes of Political factor theme are political will, diaspora, authority, occupation, corruption, external countries and cooperation with private corporations while the associated sub-themes of Capability are Infrastructure, Data Protection, HR, Tech, People, and Finance. In addition to that, the associated sub-themes of Trust are Endorsement from Officials, campaign, IT Literacy, Education, Generational Divide and Oversight/Assessment.

In terms of understanding the contextual factors affecting the implementation of e-voting in Palestine, these codes present a hierarchy in which political will is the primary contextual issues affecting the implementation of an e-voting service in Palestine, followed by capability, trust, and acceptance. Essentially, if the political will is there, then all factors can be put in place, but if the political will is not there, then it is moot to determine capabilities, trust, or acceptance. Although these four selective codes are presented separately below, it is important to acknowledge the

Fig. 2 Diagrammatical
depiction of four core themes

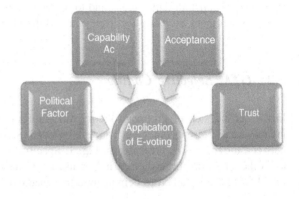

intimate relationships between them (demonstrated in Fig. 1) and that they cannot be clearly separated in theory. For instance, capability, trust, and acceptance are all both impacted by and a factor in political will.

The above-mentioned concept map can be summarized as a diagrammatical depiction of the core themes in addressing e-voting in Palestine (Fig. 2).

6 Findings

6.1 Political Will

Themes and sub-themes that contributed to the conceptualization of the core theme of political will for e-voting include current government usage of e-technology, cooperation with the private sector, authority, occupation, external countries, diaspora, and political corruption. There are essentially four actors to consider within those core theme in the implementation of e-voting in Palestine: the government, the citizens, the private sector, and external countries. In terms of political will, support must be received from the citizens, but it must first come from the government. Political will is thus the foremost issue of consideration in the establishment of e-voting. It is seen by interviews as both what is needed to implement the project and what threatens the project. Political will must have parameters. Such will must be enough to establish the need and support for the project, but not so much that it influences the project.

Authority posed a further concern, but also presented itself as a multi-faceted concept. Authority, in the context of this research, can refer to (1) the disinterest of politicians in losing their authority which yields a lack of political will, (2) current authority relating to decision-making, and (3) proposed authority relating to implementation and control over e-voting. The most frequent reason for the lack of political will was that political will must come from those in authority, but that in supporting such a system of e-voting, these officials would lose authority. Simply put, even if politicians had the political will to implement e-voting, interviewees were concerned about the great possibly of corruption. *"The thing is that the entire election process is dysfunctional even if the system were there. You know the political situation... Why will they have doubts? It is because we doubt the voting process or that the person who will take out the results will cheat."*—The Manager of Mada Internet Services Company in Gaza.

While there were examples of e-governance used in Palestine, most interviewees were not knowledgeable about these technologies and services unless directly related to their implementation. Most examples of e-services were related to banking or other private sector services. Interviewees were asked about the current state of an electronic signature in Palestine and most believed that it has not yet been developed, but that there is the ability to develop such a system. A private sector contractor working on government projects argued that it was impossible to have electronic signatures in Palestine because "e-signature could only come after the stage of building e-governance and giving it trust." This point will be further discussed in the final section of this report.

Much of the political will that does exist is driven by the desire and need to include the Palestinian diaspora. Currently, nearly 10 million Palestinians live outside the country, while only 1.7 million live in the country. E-voting is a proposed solution for including these individuals in the democratic process. When discussed benefits and the impact of e-voting, the inclusion of these external populations was a common theme. For instance, An IT specialist working for a non-profit organization, stated, "*the pros are that it will encompass all Palestinians and enable them to vote and resolve the borders and geographical issues in general and would result in every Palestinian exercising their right to vote.*" Many interviewees noted that in the current environment many Palestinians are unable to reach polling stations and e-voting would address this barrier. While some noted this barrier was related to the spread of the population, others noted that it also applies to sick and handicapped people. In fact, most interviewees saw the inclusion of the diaspora as the main benefit of e-voting. The former head of IT department of Al-Azhar University, for instance, stated that "*the benefit is that it's the only available political solution for holding elections that include the whole Palestinian population.*" The former ICT minster adds that "*e-voting would be more democratic people of this greater inclusion*". The Director of E-Government adds that "*e-voting would give voters more privacy as they would not have to vote under pressure and they could even cast their vote at night and from the comfort of their home, while the polling station atmosphere would be more stressful*".

6.2 Capability

The capability was considered in terms of infrastructure, personnel, finances, and the ability to protect the systems against fraud and corruption. Interviewees overwhelmingly agreed that Palestine has the infrastructure and human resources capability to implement an e-voting system. While a few interviewees felt that there are financial barriers, some stated that such a system would be less costly than traditional ballot methods. Interviewees almost unanimously agreed that political will trumps capability in terms of barriers to an e-voting system. The only concern that was consistently raised was related to a generational divide in the IT literacy of citizens. While interviewees agreed that Palestinians are extremely technology literature, there was a general concern for the most elderly population and their ability to use technologies. The interviewees overwhelmingly agreed that Palestine has the infrastructure to support e-voting. The Manager of Mada Internet Services Company in Gaza, for instance, stated, "*the Internet reaches over 90% of the Palestinian areas, and we have no problems. The average speed is 2 megabytes and is to become 4 megabytes as the least speed. The people are developing along.*" Similarly, the Director of E-Government described Palestine as having complete maturity in this matter. Interviewees also acknowledge that e-signatures would need to be in place in addition to the infrastructure and this has yet to occur. Similar, the strong majority of interviewees felt that Palestine has the HR capability in terms of individuals qualified to design and implement the system from a technical standpoint. One interviewee, a constitutional Judge, felt that the infrastructure nor the HR capability were, in fact, not in place to implement e-voting—"*In Palestine, there is a need to upgrade our equipment and requalify public sector workers with specific educational course in handling technology as many employees in the government sector do not have a good background in dealing with technology and the requirements of an electronic government.*" However, constitutional Judge noted that there are individuals that are well qualified and many studies abroad and then return. Interviewees expressed mixed feelings on the capability to protect the data produced through e-voting. Sentiments fell into four categories: Yes, this is a problem, and it will prevent e-voting for effectively occurring; Yes, this is a problem, but it exists will all e-services; Yes, this is a problem, but it exists will all elections; and No, this is not a problem. Regarding the ability to protect e-voting data, comparisons were often made to the ability of banks to protect sensitive data and to interact with customers electronically. Banks can verify identification and therefore, serve as an example of how individuals are already using and trusting e-services.

6.3   Trust

There are, in fact, many issues raised within the interviews regarding trust. There are concerns over hacking, fraud by citizens, and corruption by political leaders. The consensus among interviewees was that trust could accomplish through efforts such as gradual implementation, education and social awareness, the support of respected leaders, and a generational shift. Further efforts to improve trust issues include proper oversight and assessment. While some interviewees feel this should occur by a neutral third party in the private sector, others felt that it should occur through a government commission. Interviewees noted that finding a neutral assessor could prove problematic given the state of Palestinian politics. There was also fear that there would be a conflict over management of the project that could limit its implementation. Relating back to political will, many interviewees felt that politicians are simply not comfortable with e-voting, because as a Private Sector Contractor for Government stated, "*its transparency will expose them.*" This indicates that there is a lack of track for politicians and the political system.

The interviewees broadly felt that full trust of the system could not be established. Rather, it was a factor creating more trust in an e-voting system that exists for comparable or traditional systems. Much of the limitation in trust stems from the human aspects of the technology. For instance, an interviewed lecturer in Information Systems at the University of Greenwich exclaimed that "a system that is 100% protected does not exist because anything that is made by humans is also vulnerable to infiltration by other human beings." This refers to infiltration in multiple aspects, such as identity fraud, hacking into the system both domestically and by foreign entities, and by party corruption. If the voters do not perceive that their vote will be counted democratically and reliably, they will not trust the system as a whole. The Director General of Information Technology stated, "The matter [of trust] depends on whether we trust those who oversee it and their integrity. If there are signs of corruption in their daily lives, how would you expect to have faith in them handling the system." Trust, therefore, must occur at multiple stages (design, implementation, reporting, assessment) and across many actors to be effective.

6.4 Acceptance

For e-governance to be accepted, all of the factors have to be in place. There has to be the political will, the capability, and the trust. If Palestine lacks any of these factors, the project will not be accepted by the citizens or the leaders.

"Issues of development do not usually come about because of a single need, but due to many different needs."—A Private Sector Contractor for Government

The interviewees discussed methods for promoting acceptance of the system, such as campaigns to promote IT literacy, educate the citizens on services available, and to implement systems gradually. "In my opinion, e-voting should be offered as a complementary alternative to traditional ways of voting; not as a replacement. If the parties and the candidates do not support the idea, they will also encourage their constituents not to use it. They could tell them this could cause a manipulation of the results, and therefore they would reject its use and convince their supporters not to use it either. This would prevent reaching the required results."—Senior Lecturer in Information Systems at the University of Greenwich.

As previously addressed, a key component to promoting acceptance is establishing trust through oversight. The Director General of Information Technology stated, "if those overseeing elections presented themselves to the public and were people of integrity, the people will trust." Interviewees had mixed interpretations of the impact of an e-voting system. Nearly all interviewees believed that electronic voting would result in increased participation in elections because it would enable people vote without pressure and even encourage many individuals who have not previously voted and handicapped individuals who were previously unable to get to polling stations. Moreover, the ability to allow the diaspora to vote was an essential component of support in the overall system, but also the belief that e-voting would increase voter turnout. Regarding the ability of e-voting to increase voting rates, a private sector contractor working with the government, stated, "to a very large extent. Especially since the Palestinians living abroad will be given the opportunity to participate, and the youth in Gaza will particularly encourage the idea." Another interviewee noted that the increase in participation would come about eventually, but not immediately unless political will changes. Only one interviewee—Unit Head of Networks and Security at Palestinian Telecom Company— predicted poor participation: "*I foresee poor participation... because they are used to the classical voting method by writing down his vote and inserting the vote in a box. That is the Palestinians' ideal on voting and is yet to change.*"

7 Discussion

Findings on the contextual issues affecting the implementation of an e-voting service in Palestine have herein been presented. GTM revealed four core themes of factors surrounding perceptions of e-voting implementation in Palestine as expressed by nineteen elites: political will, capability, trust, and acceptance. The data reveal that implementation is a complex phenomenon and is both impacted by political, cultural, and organization factors [1, 3]. The interview data reveal that Palestine is, indeed, a complex political environment and that this complexity is a barrier to the implementation for e-voting, but the complexity also creates a greater need for implementation of e-voting. Given this great need and the perception of most respondents that e-voting would increase participation, e-voting would improve democracy in Palestine. Perhaps one approach to implementation starts with other areas of e-democracy. "*There are massive opportunities for the enhancement of democratic processes* via *electronic interaction. It is perceived as an enabler for the enhancement of existing democratic practices as well being a catalyst for democratic transformation*". It can be argued that the ICT can not only enhance existing democratic practices but also can pave the way for democratic transformation by means of the direct participation of all citizens. Thus two ideas emerge with respect to the application of ICT to democracy. One is based on the "presumption that any political use of new technologies takes place within existing institutional frameworks of parliaments, executive branches, and political parties" [37], and the other advocates the transformation of representative forms of democracy into more direct forms [38]. Furthermore, most research shows that various government initiatives employ the Internet and ICT to provide services and information. In the case of Palestine, therefore, the e-government can start in other sectors. This will likely increase trust and participation. However, it will not be as easy to address issues of trust as they relate to corruption and control of the process.

Trust emerged within this research as a core theme and issue of concern. This finding aligns with the literature and has proved to have a particular significance in developing countries. Several studies have examined the issue of trust in relation to electronic elections. In Brazil, a study revealed a great deal of evidence that the e-voting system was widely used and appreciated, and that the election results it produced were, with few exceptions, accepted to be the correct aggregate of individual citizens' actual votes. However, the formation of plausible cause-and-effect relationships between a technical/institutional arrangement and trust, as expressed in citizens' opinion surveys and demonstrated by their actions, posed a theoretical challenge [39].

The case that is likely the closest to Palestine in regards to the implementation of e-voting is Iran. The comparison can be drawn from the findings presented herein on Palestine to the Iranian case, particularly in discussing the role of trust. It is important to note, therefore, that in Iran, there have been some cases where electronic voting was used and achieved a good level of success, and there are other times where 'trust' issues prevented further implementation of electronic voting. The main reason relates to the lack of trust in the accuracy and correctness of software [40].

8 Conclusion

The findings presented herein are both supported by the extant literature and ground-breaking in the depth they present. As the findings align with research conducted in other countries and regions regarding the challenges of implementation and their relation to factors such as the complex political factors [1, 2, 8, 41, 42], the research has also advanced the literature in its focus on the unique environment of Palestine. Not only is Palestine unique in its need for e-voting for incorporating the mass diaspora into e-democracy, but it is also unique is the details of the challenges that are faced in implementation, as the data have revealed. However, the challenges more broadly are represented of those that have been presented in case studies of other countries. For instance, while technological capabilities and trust of both citizenry and officials are known factors associated with e-government and more specifically e-voting [10, 11], this research has provided greater depth to these issues in the specific context of Palestine, such as the impact of the occupation, political will, political history, and diaspora as both a need for e-voting and an obstacle to implementing e-voting. In doing so, this research has achieved the three purpose areas outlined as motivations for the study. This research concludes with the charge for future research to saturate the above-mentioned categories by gathering more in-depth qualitative data which will drive this research to form grounded theory related to the research question.

References

1. McGrath, K., Maiye, A.: The role of institutions in ICT innovation: learning from interventions in a Nigerian e-government initiative. Inf. Technol. Dev. 16, 260–278 (2010)
2. Alshawi, S., Alalwany, H.: E-government evaluation: citizen's perspective in developing countries. Inf. Technol. Dev. 15, 193–208 (2009)
3. Luo, G.: E-government, people and social change: a case study in China. Electron. J. Inf. Syst. Dev. Ctries. 38 (2009)
4. Chen, Y.N., Chen, H.M., Huang, W., Ching, R.K.H.: E-government strategies in developed and developing countries: an implementation framework and case study. J. Glob. Inf. Manag. 14, 23–46 (2006)

5. Dada, D.: The failure of e-government in developing countries: a literature review. Electron. J. Inf. Syst. Dev. Ctries. **26** (2006)
6. Heeks, R.: Most eGovernment-for-Development Projects Fail: How Can Risks be Reduced?. Institute for Development Policy and Management, University of Manchester, Manchester (2003)
7. Tan, M.: An investigation of e-government services in China. Electron. J. Inf. Syst. Dev. Ctries. **57** (2013)
8. Rodríguez Bolívar, M.P., Alcaide Muñoz, L., López Hernández, A.M.: Scientometric study of the progress and development of e-government research during the period 2000–2012. Inf. Technol. Dev. 1–39 (2014)
9. Heeks, R.: e-Government in Africa: promise and practice. Inf. Polity **7**, 97–114 (2002)
10. Gilbert, D., Balestrini, P., Littleboy, D.: Barriers and benefits in the adoption of e-government. Int. J. Public Sect. Manag. **17**, 286–301 (2004)
11. Irani, Z., Kamal, M., Al-Sobhi, F., Weerakkody, V., Mustafa Kamal, M.: An exploratory study on the role of intermediaries in delivering public services in Madinah City: case of Saudi Arabia. Transform. Gov. People Process Policy **4**, 14–36 (2010)
12. Ndou, V.: E-government for developing countries: opportunities and challenges. Electron. J. Inf. Syst. Dev. Ctries. **18** (2004)
13. Stoltzfus, K.: Motivations for implementing e-government: an investigation of the global phenomenon. In: Proceedings of the 2005 National Conference on Digital Government Research, pp. 333–338. Digital Government Society of North America (2005)
14. Basu, S.: E-government and developing countries: an overview. Int. Rev. Law Comput. Technol. **18**, 109–132 (2004)
15. Shafi, A.-S., Weerakkody, V.: Understanding citizens' behavioural intention in the adoption of e-government services in the state of Qatar. In: ECIS, pp. 1618–1629 (2009)
16. Gil-García, J.R., Pardo, T.A.: E-government success factors: mapping practical tools to theoretical foundations. Gov. Inf. Q. **22**, 187–216 (2005)
17. Yildiz, M.: E-government research: reviewing the literature, limitations, and ways forward. Gov. Inf. Q. **24**, 646–665 (2007)
18. Rössler, T., Leitold, H., Posch, R.: E-voting: a scalable approach using XML and hardware security modules. In: Proceedings of the 2005 IEEE International Conference on e-Technology, e-Commerce and e-Service, 2005, EEE'05, pp. 480–485. IEEE (2005)
19. Watson, R.T., Mundy, B.: A strategic perspective of electronic democracy. Commun. ACM **44**, 27–30 (2001)
20. Heeks, R., Bailur, S.: Analyzing e-government research: perspectives, philosophies, theories, methods, and practice. Gov. Inf. Q. **24**, 243–265 (2007)
21. Musso, J., Weare, C., Hale, M.: Designing Web technologies for local governance reform: good management or good democracy? Polit. Commun. **17**, 1–19 (2000)
22. MAAN: تعقيب لجنة الانتخابات حول التسجيل والتصويت الالكتروني, http://www.maannews.net/Content.aspx?id=632707
23. Al-amer, M.A.: Electronic democracy strategy for Bahrain (2009)
24. Graham, T., Witschge, T.: In search of online deliberation: towards a new method for examining the quality of online discussions. Commun. AUGUSTIN THEN BERLIN- **28**, 173–204 (2003)
25. Svensson, J., Leenes, R.: E-voting in Europe: divergent democratic practice. Inf. Polity. **8**, 3–15 (2003)
26. Campaign, P.N.C.R.: http://www.pncregcampaign.org/, http://www.pncregcampaign.org/
27. Shat, F.J.F., Mousavi, A., Pimenidis, E.: Electronic government enactment in a small developing country—the Palestinian authority's policy and practice. Commun. Comput. Inf. Sci. **441**, 83–92 (2014)
28. Bryan, C., Tsagarousianou, R., Tambini, D.: Electronic democracy and the civic networking movement in context. Cyberdemocracy Technol. Cities Civ. Netw. 1–17 (1998)
29. Campaign, P.N.C.R.: PNC registration machine ready for voter registration

30. Kiayias, A., Zacharias, T., Zhang, B.: End-to-end verifiable elections in the standard model. In: Advances in Cryptology-EUROCRYPT 2015, pp. 468–498. Springer (2015)
31. Ntaliani, M., Costopoulou, C., Karetsos, S., Molhanec, M.: Citizen e-Empowerment in Greek and Czech municipalities. In: E-Democracy–Citizen Rights in the World of the New Computing Paradigms, pp. 124–133. Springer (2015)
32. Myers, M.D., Avison, D.: Qualitative Research in Information Systems: A Reader. Sage (2002)
33. Silverman, D.: Qualitative Research. Sage (2010)
34. Urquhart, C.: Grounded Theory for Qualitative Research: A Practical Guide. Sage (2012)
35. Myers, M.D.: Qualitative research in information systems. Manag. Inf. Syst. Q. **21**, 241–242 (1997)
36. Glaser, B.G.: Conceptualization: on theory and theorizing using grounded theory. Int. J. Qual. Methods **1**, 23–38 (2002)
37. Gibson, R., Römmele, A., Ward, S.: Electronic Democracy: Mobilisation, Organisation and Participation Via New ICTs. Routledge (2004)
38. Westen, T.: E-democracy: ready or not, here it comes. Natl. Civ. Rev. **89**, 217–228 (2000)
39. Avgerou, C.: Information systems in developing countries: a critical research review. J. Inf. Technol. **23**, 133–146 (2008)
40. Kahani, M.: Experiencing small-scale E-Democracy in Iran. Electron. J. Inf. Syst. Dev. Ctries. **22**, (2005)
41. Alvarez, R.M., Katz, G., Pomares, J.: The impact of new technologies on voter confidence in Latin America: evidence from e-voting experiments in Argentina and Colombia. J. Inf. Technol. Polit. **8**, 199–217 (2011)
42. Walsham, G.: ICTs for the broader development of India: an analysis of the literature. Electron. J. Inf. Syst. Dev. Ctries. **41** (2010)

The Perceived Impact of the Agile Development and Project Management Method Scrum on Process Transparency in Information Systems Development

Karlheinz Kautz, Thomas Heide Johansen and Andreas Uldahl

1 Introduction

Over the last decade agile information systems and software development (ISD) has received much attention from researchers and practitioners as an approach for dealing with change and the unpredictable and hardly controllable elements of ISD in a dynamic environment. While numerous publications claim a positive impact of agile development and in particular Scrum on ISD, only little empirical work exists to verify these claims. The literature review, which was part of this study, uncovered some notable exceptions. To further contribute to this body of knowledge we set out to answer the following two research questions: What impact has the introduction of the agile development and project management method Scrum on process transparency in ISD? What is the effect of any deviations from the guidelines for Scrum? The results we present in the following are part of a larger project where we developed a framework comprising the six concepts productivity, quality, team leadership, employee satisfaction, process transparency, and customer satisfaction for the investigation of the impact of Scrum [1]. As ISD has long been understood as a social process with an acknowledged importance of social

A prior version of this paper has been published in the ISD2016 Proceedings (http://aisel.aisnet. org/isd2014/proceedings2016).

K. Kautz (✉)
College of Business, RMIT University, Melbourne, Australia
e-mail: kautz@uow.edu.au

T.H. Johansen
Progressive A/S, Herlev, Denmark
e-mail: thj@progressive.dk

A. Uldahl
Ernest & Young, Søborg, Denmark
e-mail: Andreas.Uldahl@dk.ey.com

© Springer International Publishing Switzerland 2017
J. Gołuchowski et al. (eds.), *Complexity in Information Systems Development*,
Lecture Notes in Information Systems and Organisation 22,
DOI 10.1007/978-3-319-52593-8_15

237

interaction [2–4] in this paper we concentrate on one of these concepts which is explicitly related to social interaction, namely Scrum's impact on process transparency in ISD. In the remainder of the paper we first briefly introduce Scrum, and then we describe our theoretical background and the research setting and method. Subsequently we present and discuss our research results against the existing literature on Scrum and relate them to our earlier findings concerning the other concepts of our framework which were explained in more detail in [5–7] as well as to complex adaptive systems (CAS) theory, a theory which is considered to provide a theoretical foundation for ISD [8] and in particular agile development [9]. We finish with some conclusions and an outlook to future research.

2 Scrum—An Agile Method

Scrum is an agile ISD method with a strong focus on project management, which was formalized and tested by Schwaber and Sutherland in the mid-1990s [10, 11]. Scrum focuses on an iterative and nimble development process, emphasizing among others transparency, both in the sense of clarity and visibility and a cooperative, collegial leadership style and cooperation in and between the development team and the customers. In Scrum the development team is called the Scrum team. Unlike traditional development projects where analysts, developers and testers are typically separated, Scrum teams are built on an interdisciplinary basis and comprise all these roles in one team preferably in one physical location. This structure, as well as Scrum's focus on self-organization aims at creating team dynamics and a better understanding of the tasks to be performed jointly. Internally, the role of the Scrum master will provide leadership, motivate and facilitate the team in line with the Scrum values, practices and development process. The role of the Product owner has the responsibility to represent the project and product externally to other stakeholders and customers and to handle and manage the tasks that appear in the product and release backlogs [11]. A Scrum development process is structured through a product backlog, which is a prioritized list of required business and technical functions of the envisioned product. It might change in line with the customer's new needs. A release backlog is a prioritized subset of the total product backlog and defines the functions to be included in a release. A Scrum, performed in so-called sprints, is a set of development tasks and processes which a Scrum team carries out to achieve a given sprint goal. The length of a sprint is predefined. It typically lasts between 5 and 30 calendar days [11]. What needs to be done during a sprint is determined by a prioritized sprint backlog, which is determined together with a sprint goal before the start of each sprint by the team and Scrum master and others, if necessary, at a planning meeting. The work process during a sprint is visualized on a public Scrum board which contains information about task status such as not started, in progress, finished. Throughout a project a Burn down chart shows the amount of work left to do versus time over a given period [10]. In short daily Scrum meetings project members briefly present what they have done during the preceding day, which tasks they take on that day, as well as any challenges and obstacles that might have prevented them from carrying out their work without any solution being discussed. Scrums of scrums are additional short meetings by the Scrum masters of projects, which consist of several Scrum teams. At the end of a sprint a sprint review meeting takes place where the Scrum team, the Product owner, other management, and one or more representatives from the customer [11] assess the team's development process and progress in relation to the predefined sprint goal. Finally the Scrum team, the Scrum master and possibly the Product owner hold a meeting, called a retrospective, to secure learning and further improvement in the team where both the process and the product are assessed and discussed by each individual team member.

3   Literature Review and Theoretical Background

In our study we were interested in the impact of a specific method, namely Scrum on ISD. Our literature review was therefore focused on that particular approach and not in general on project management methods' or agile methods' impact on ISD. This limited our sources to writings which take their starting point in agile software development. We combined a concept-centric with an author-based approach [12] and applied backward referencing of sources. Our search with keywords such as 'impact of Scrum', 'effect of Scrum', 'impact of Scrum implementation', and 'effect of Scrum implementation' primarily in Google, Google Scholar and IEEE sources lead to about 90 sources of which 8 dealt more precisely with our research problem. An additional 8 sources were identified through the other mechanisms. From that literature we derived a number of concepts and for these concepts indicators for the impact of Scrum on ISD processes and projects. The resulting framework consisted of the identified, interrelated concepts, process transparency, team leadership, productivity, quality, employee satisfaction, as well as customer satisfaction and 38 indicators, which defined the concepts on a more detailed level. Here we are focusing on Scrum's impact on the first concept. We have reported Scrum's impact on the other concepts elsewhere [5–7]. Process transparency refers to one of the areas that distinguish Scrum noticeably from the typical waterfall model development processes. It is argued that Scrum only works as intended if the whole process is at every point in time transparent both in the sense of clarity and openness as well as in the sense of visibility for everyone involved including those in leadership roles, the developers and the customers [10, 11]. This view is shared by Moe and Dingsøyr [13] who emphasize the relationship between process transparency and team collaboration and effectiveness. Schwaber and Beedle [11] in this context stress the effect visibility has on conflict resolution when disagreements or misunderstandings arise in teams. These authors agree that practices such as sprints, different kinds of regular planning, information and review meetings as well as artefacts such as Scrum boards, Product backlogs and Burn down charts promote visibility and provide an overview and a status of the ongoing work. Sutherland and Altman [14] also make a case for the impact enhanced visibility of task and process status has on developers and their collaboration as they as members of development teams are constantly aware of their own and others tasks and responsibilities and thus are able to reduce waste time. We use these sources to investigate the indicators task status and overview and team collaboration. Another emphasis related to transparency is put on the planning activities, in particular the introduction of an iterative, participative and frequently applied estimation process and the connected reliability and credibility of the resulting estimates [11]. We adopted both estimate credibility and estimation process as indicators for process transparency. Finally, we put specific weight on the relationship between customer involvement, customer satisfaction and process transparency as customer involvement both requires transparency, but also supports it. We therefore decided to include customer involvement as a fifth indicator for process transparency in our investigation.

4   Research Setting and Method

We chose a case study approach to research the impact of Scrum on ISD processes and projects. The chosen case organization has approximately 40 years of experience in solving complex IT tasks. Some years ago it changed from being publically owned to private company. It has about 3000 employees, who are involved in the development of administrative and statutory software solutions. The investigated case department falls into the latter category and has 45 employees. Its sole product is a case management system for municipal job centers, which gives administrators the opportunity to work across different platforms. For the development of the case management system, the department previously followed the traditional waterfall model. In 2011 it launched the implementation of Scrum as the preferred development model. At the time of our investigation in 2012, the department had completed three full releases with the use of Scrum. As such the department had the profile of the unit of analysis we were looking for, an organization that had recently, within the past year, chosen to implement Scrum, and that had previously used the traditional waterfall model. With the former model still in their minds we expected the employees to make candid assessments of the impact of Scrum as compared to the past. As we were not able to neither make direct observations concerning task status and overview, team collaboration or estimation process nor measurements such as number of encountered and resolved conflicts or tasks finished on estimated time, etc., we chose to directly ask respondents about their perceptions of the given concepts. The indicators, which we had derived from the literature review, were therefore transformed into direct questions for our interviews, which we validated with 2 employees in a small pilot study before putting them to the 11 interview partners who were available for the study. We developed 3 largely overlapping interview guides for the three stakeholder groups, with 6 developers as respondents, 4 respondents in leadership roles such as Scrum master, Product owner or unit managers and one representative from the service department, which is responsible for customer liaisons. All interviews were recorded, transcribed and handed over to the respondents for approval. The results of our analysis were also presented to the participants of this study and the case organization at large.

The data collection with standardized interviews allowed both collections of qualitative and quantitative data. We first asked the respondents to numerically assess, on a scale from −5 to +5, for each indicator its individual change, improvement or decline, as compared to the situation before the implementation of Scrum and then to evaluate its impact on the concept in question, here process transparency. After that quantitative judgment we asked into the reasons for these assessments, which provided rich qualitative data. This combination of data allowed for data and method triangulation to improve the validity of our findings [15]. The subsequent analysis was based on mean values for the quantitative data within each indicator; these were interpreted on the basis of the qualitative opinions. The results for the individual concepts were then compared and discussed with regard to published Scrum guidelines, findings from the literature, their relation to each other and to CAS theory. It is worth pointing out that the numerical element of the collected data should be considered secondary. The interviews were intended as the primary source to collect qualitative data with a statistical element—and not vice versa. The quantitative data was exclusively used to create an indication and an overview over any specific area.

5 Findings—Scrum's Impact on Process Transparency

Table 1 summarizes the respondents' assessment on Scrum's impact on process transparency.

Table 1 Scrum's impact on process transparency

	Improvement	Impact on process transparency	Range of score in both dimensions
Task status and overview	0.4	0.3	−5–5
Team collaboration	2.9	2.1	−1–5
Estimate credibility	1.8	1.4	−2–4
Estimation process	3.0	1.5	0–3
Customer involvement	0.25	0.25	0–2

5.1 Task Status and Overview

There was considerable disagreement among the respondents as to whether there had been an improvement or a deterioration concerning this indicator as the responses on both dimensions ranged from −5 to 5 on the scale. None of the respondents perceived this parameter as unchanged and everyone had an opinion. The answers were distributed more or less into two factions, one in which both dimensions were perceived exceedingly positive, and one where both were considered highly negative. A number of different reasons were provided for the negative assessments; however one issue was mentioned several times, namely that while the overview over the tasks and features that individuals worked on within the smaller sub teams had been improved, the general overview over the status of the whole actual release had decreased. The following responses illustrate more specifically what the respondents said about this issue:

> (…) Well, there may be a next delivery, it may consist of two PBIs [Product backlog items], or it may consist of 20 PBIs. There is not a chance to keep track. I go and check in the product backlog, but I have not found any system in the Product backlog yet to see it. And when I ask the others, then there is no one else who knows it (…).

> (…) it is much harder now to get an overview of what is going on in the individual teams regarding the PBIs because there are many more small clumps, many more clumps across the teams. Unless you go to all Scrum meetings and you can actually figure out our very inextricable Product backlog, then it did not become easier. (…) or you have to walk down to the POs [Product owners] and look at their board, which is not very clear, so I do not think it has become easier.

> (…) before Scrum I would say that it seemed as if we had a better overview over what actually was going on, and we do not have that now, because now there are the teams which sit and work with small PBIs or something like that, so it has just become worse (…).

The decomposition of functions into deliveries, sub-deliveries, and ultimately into Product backlog items had its negative sides both for some developers and leaders. The tasks had become small and were perceived as very confined so that overview and track were lost. It was possible to get an overview by looking into the Product backlog, but apparently this was such an arduous process—the Product backlog was perceived as poorly organized and implemented—that this was omitted by most respondents with the consequence that they lost their overall view of the task status. The following statement is worth highlighting as it places its somewhat negative assessment into the context of the new work practices:

> (…) It's more because now that you sit very good here with a PBI, and you really do not have to be concerned about what all the others (…) are doing, the less it is something which interacts with your area, but often when we are approaching a release, I'm actually quite curious and wonder what the others have done, because usually there is not that much time, and I only hear about it when we have our weekly meetings.

This respondent expressed that it actually was not necessary to know what the developers were doing, but that his unsatisfied curiosity still had the effect that he sometimes missed the lacking overview. Those who provided positive ratings emphasized that previously there had been no methods or tools to create an overview. Several of them also agreed that the decomposition of tasks had improved the general overview over the ongoing work significantly. The breakdown of tasks was thus perceived both as positive and negative, as it also had been blamed for a lack of overview, but for some respondents this was compensated for by Scrum's aptitude and capability to visualize things in physical form which was stated as one of the most important factors to establish an overview over the task status and thus contributed to process transparency:

> Well it is because some of the artifacts that come with Scrum, they get things out into the open; there is the one that you have the product backlog items, that you have bought into for the actual sprint are on the Scrum board. (...) And then there is that one that you get information out of the computers and onto the walls, this practice has just spread so much more than when we had the waterfall approach, so that's what I think.

5.2 Team Collaboration

The responses on both dimensions regarding team collaboration ranged from −1 to 5 on the scale. This indicator showed the largest positive change in perception. The responses spanned in both dimensions from a slight decline to a large number of positive assessments from the respondents both across their job positions and their professional background with almost everyone agreeing to this positive development. One of the most important and most mentioned reasons for the improvement was the new way employees were physically located:

> (...) And that after we in the Scrum teams have been moved around, so that now we sit together in a team, and previously one sat with one's own profession, and now one has an integrated team, and is pulled together and we get to know each other much better.

Several respondents also expressed that previously there had been no collaboration across the different professional disciplines, which resulted in that this change was perceived as even more significant. As such cooperation had not existed in the past, this change made a strong impression on the individual employees. The respondents generally agreed that it had provided a broader perspective and a better overview—one even expressed that he had learned to read code:

> (...) I know so much now about what we do, I can almost read the code myself because I have stood next to the developers so many times. I would not say I ever get around to program anything, but just to understand what it says on the sheets and the screen written in red and green, that is fantastic, isn't it.

This respondent ended up rating this indicator with 5 on both dimensions. A single respondent assessed the change negatively with −1. The expressed opinion and reasons however, indicate that the respondent had interpreted the question differently than the others:

(...) Of course, it just so happens that when you are not sitting together any more, it gets worse. Socially, it has no effect even if you are not sitting together with the other 10 who work at the backend anymore (...).

Although this respondent's assessment could or should not be compared with the other respondents due to the possible misinterpretation, in itself this opinion is interesting. It took an angle of the change that none of the other respondents had considered. Where the other respondents assessed whether cooperation across professional groups had improved, this respondent assessed how cooperation within the professional disciplines had changed, which for this employee apparently had not been a positive development.

5.3 Estimate Credibility

The responses concerning estimate credibility on both dimensions ranged from −2 to 4 on the scale. During our interviews a tendency quickly became apparent concerning this indicator. The respondents mostly agreed that two actions had been the cause for an improvement of the estimates' credibility. These two actions were (1) the change of the scope of the estimation process, and (2) the change of who was responsible for the estimates. The following three quotes are reflections of the first measure:

Probably it is the more defined and delimited process which is easier to estimate. So before we talked, you know, about 600, 1200 and 2000 h. You can't really do that, and that was also a guesstimate for most of us, but that's often what happens when one has to estimate, but now it is, (...) we are much closer to estimate correctly. So I will say it has improved.

It [the task] has been broken down, so now it is possible to estimate correctly. When there are small estimates, there is a difference whether you estimate something over half a year or for 14 days, as we run sprints in the moment, so it's easier to get it right.

It has become much better because the estimates are regularly revisited, in the old days some analysts and some managers estimated something and then it was simply sent through the waterfall, nothing was ever revisited or revised; now estimations take place all the time.

Based on these opinions, we can conclude that Scrum's iterative workflow, together with a reduction of sprints to 14 days helped to eliminate or delimit some of the uncertainties which typically influence and impede estimates. The second measure regarding the change of responsibility for the estimations is reflected in the following respondent's opinion:

I do not really think that estimates were actually made before. I do not think so. Well, I think there were some leaders who sat down and made some estimates. And it was a bit those that one followed. But now that it is those directly involved who estimate, it is obvious that estimate credibility has risen. So it has a positive effect on the estimates.

Assigning the responsibility for the estimates to the Scrum teams, according to this respondent, was the cause of increased credibility. A third point of view came from another respondent. This developer looked differently at the issue and thought that the estimates' credibility had decreased. This however was considered more of a temporary challenge as he reasoned that that was the case because the team itself was still quite inexperienced with regard to estimation and therefore had not yet got used to the process.

5.4 Estimation Process

The estimation process had been identified as an independent indicator. To allow for a comparison the question concerning this indicator had only been posed to the respondents who held a leadership role as it were those who had performed the estimates in the past. The resulting responses on both dimensions ranged from 0 to 3 on the scale and pointed towards a significant improvement, however with a relatively small effect on process transparency. The following explanation given by one of the respondents represents the general opinion provided by this group:

> It is much easier than before, that is certainly a +3, I think, but then it is still really quite cumbersome and difficult, we have improved, but it is still not good.

When asked directly whether the improvement could be attributed to the implementation of Scrum the response was:

> Yes absolutely, but as we now have entered the Scrum arena, we are definitely not benchmarked as the best in class, we certainly can get much better.

These two answers indicate that the respondent had a feeling that the estimation process had improved, but that there was still room for enhancements before the case organization could compare itself with more successful and well-functioning Scrum teams in other organizations. The reason that the effect on process transparency was perceived as less significant than the actual improvement was as the respondent, in line with the critical developer referred to above, emphasized that the case organization was still quite inexperienced with the new estimation process and therefore still in a learning process in which they gradually, from sprint to sprint, adjusted the way they used Scrum in general and the estimation process in particular.

5.5 Customer Involvement

This indicator differed from the others in that almost all respondents indicated no change and answered 0 for both dimensions. This was not unexpected as the pilot interviews had pointed to that there had been little development in this area. As customer involvement is however a cornerstone of Scrum, it merits a closer investigation and the fact that the situation was nearly unchanged is in itself

K. Kautz et al.

interesting. Despite the fact that almost all respondents answered 0 on both dimensions, there was a certain direction expressed in their opinions:

> (...) I can just say that it is my impression that there might be more [customer involvement]. But I do not think that it has something to do with Scrum, I think it is becoming more a general practice.

> I have no great insight into that area, but we have more focus on it today, but it has nothing to do with Scrum, that's what my immediate thought is (...).

Thus, even if respondents agreed that there had been no change to be reported, they noted an increased focus on customer involvement as the organization apparently had recognized the importance of customer involvement and had started to provide the resources to do so. That the two above quoted respondents stressed, that this was not due to Scrum, about due to a general progression to more direct customer involvement, is still noteworthy.

Another remarkable facet of this indicator in this context was that the respondent representing the service department, the only department with direct customer contact was the only one who did not reply 0 on both dimensions. Instead, the indicator is rated positive with a 2 on both dimensions; however on the grounds that customer involvement, which was still very limited and only indirect in the case unit, was expected to rise. This is the only reason why the mean values on both dimensions were not 0.

6 Discussion

The investigation of Scrum's impact on process transparency in ISD was part of a larger study which both developed and applied a comprehensive framework consisting of six related concepts. Although a presentation of the overall result would give a more comprehensive portrait of the method's impact, we focus here on process transparency as one of the key concepts. This still provides some valuable insights, which we will relate to the findings concerning the other concepts that we have presented and discussed in more detail elsewhere [5–7]. As a starting point for our discussion we summarize the results of our analysis of Scrum's impact on process transparency in the case unit.

We found that there had been an overall positive change in the respondents' perception of all indicators. However the members of the case unit were not in agreement concerning task status and overview over the functions and features which individuals, small teams and the entire development team were working on. Some thought that more transparency and a better overview was provided through the physical visualization Scrum provided through Product backlogs and Scrum boards, while others sensed that the decomposition of tasks in smaller chunks contributed to a loss of feeling a wholeness and having an overview over the entire project and product. Regarding internal team collaboration there was a broad consensus that there was a significant positive development. Most respondents provided as a reason for this the new way the employees were physically placed closely to each other in the unit. As they were sitting together in their team, they had become able to complement each other across their disciplines, which created transparency and a new and better understanding of areas outside of their own and usual responsibilities. The majority of the respondents held the opinion that the credibility of the estimates as a basis for planning and carrying out the work had improved significantly. They reasoned that this was a result of organizing the tasks in manageable sizes and shorter time periods as well as of the fact that estimation now did not involve just management, but those who actually were to carry out the work. This argument was also provided for a perceived enhancement of the estimation process as such. The estimation process was felt simpler and more appropriate, but the respondents also saw room for further improvement to come with their growing experience with the process. Regarding the involvement of customers the respondents largely agreed that no change had occurred. Just the representative of the service department who was the only respondent with contact to customers had a strong positive perception of this indicator which however was solely based on expectations for the future.

Our study is built upon subjective perceptions; as with all qualitative studies of this kind we of course have to take the danger of positive bias and a respondent's tendency of reporting future expectations rather than stating actual perceptions into account. However, the fact that the respondents reported no or only minimal impact on some of the indicators gives confidence that the reported efforts were genuine rather than showing a general positive bias. On this background, we compare our empirical data first with the literature on agile ISD and project management and in particular the identified writings about Scrum. According to these sources, there are a number of areas that impact on process transparency, these are: Customer and user involvement, Product backlog, Burn down charts, Scrum team, sprints, as well as meeting practices such as daily Scrum meetings, retrospectives meetings, sprint review meetings, and Scrum of scrums meetings.

Although we did not have access to customers we collected, to the extent possible, data about this indicator. The case unit had not utilized customer involvement as intended by the method. This was an area where there was room for improvement both with regard to involving customers in reviews, which could contribute to raising quality of deliveries, but even more so on an ongoing basis in all vital development activities. The lack of change in this area could be explained with the fact that the product was a standard solution offered for multiple municipalities with significantly different needs. Another explanation for not involving customers to a larger extent could however be that the case organization was concerned that individual incoming requests would be too diverse to be fully integrated in the standard information system under development. Another reason could simply be a lack of experience with customer involvement, a deficiency the organization intends to resolve in the future to make the most of Scrum. In any case many of the advantages customer involvement might have, were only to a limited extent and not as much as described in the literature [10] perceived by the case organization. Due to our limited data we will not pursue any customer related issues further when discussing the relationship of process transparency to the other concepts, which make up the full framework.

The positive effect of breaking down tasks into Product backlog items had not been perceived by all members of the case unit. The overview that a Product backlog creates [11] had not been conveyed well to and recognized properly by the individual developers. Instead, the smaller product backlog items had led to a decreased center of attention on the entirety of the task and project, since the developers now put more focus on their own tasks. It was therefore not surprising that some respondents felt that the expected overview over the status of the tasks as well as features and functions, which they as sub teams or as the whole development team were working on, had not achieved the desired improvement that a Product backlog could bring. The lack of the perceived adequate structure and management of the Product backlog was astonishing, as the product owners in the case organization used the possibilities to organize Product backlog items consistent with releases in their Product backlogs, which according to the literature should create process transparency. In contrast, the introduction of Burn down charts [10] unanimously had a positive effect on the task status and overview of the features, which the teams and their members were busy with. Combined with the Scrum board it provided the developers with the opportunity to see what everyone was working on in the individual sprints, and how within and beyond a sprint work progressed. The case unit had followed the guidelines from the literature and benefitted from them, which probably also compensated for the perceived problems with the Product backlog.

Furthermore the case unit followed the literature as intended regarding the composition and specifications for the expected benefits of Scrum teams [11]. The size and co-location of the teams as well as their multi-disciplinary composition created the desired dynamics and self-organization. As a result of this largely successful change the case unit enjoyed the benefits of the employees' newly acquired and enhanced mutual understanding and insight into their colleagues' work and the cooperation and team collaboration between them. The newly gained process transparency in the Scrum teams contributed to a large extent to employee satisfaction as well as to quality, in particular with regard to a significant reduction of defects per KLOCs (Kilo Lines of Code) as testing was now jointly performed by dedicated testers and developers who had roles as analysts and programmers [6]. Process transparency is also related to team leadership as a well-functioning Scrum team is the result of the Scrum master's leadership too. Such leadership allows for self-organization where everyone has insights into the other team members' tasks, while at the same time the Scrum master has a clearly defined role [11]. When there is a need for input from a specific team member, the other team members are not unnecessarily disturbed, as the tasks have been clearly defined, broken down and distributed. If in doubt, the Scrum master is available to facilitate or solve the problem. This had to a large extent been achieved in the case unit [7].

In case unit the improvements with regard to the estimation process and the credibility of the estimates, were ascribed to the defined length of the sprints, which based on shorter time periods and manageable tasks ended with executable functions, and not as in the literature [10, 11] attributed to the utilization of the Product backlog. Estimation had become easier than in the past where the estimates had to cover a time period of many months. The case unit had chosen to bring down the length of sprints to 14 days, something, which Schwaber [10] actually discourages new and inexperienced Scrum teams to do. For the case unit this however turned out to have been a good decision. Despite the fact that the case organization was still in a learning process and only slowly matured its use of Scrum, it had successfully managed to make a significant positive change with the introduction of sprints. Although not mentioned explicitly by the respondents with regard to the indicators investigating process transparency daily Scrum meetings, retrospectives and Scrum of scrums meetings have among others the objective to create transparency. Daily Scrum meetings support that everyone in a team gains insight into what the others are working on and at the same time makes it difficult to conceal modest work efforts, as employees publically have to communicate and document their results [6]. The latter had a substantial impact on employee performance in the case unit. It affected productivity positively as openly explaining why a task took longer than expected had the psychological effect that it deprived the employees of the opportunity to hide behind a task longer than necessary [6]. The increased number of meetings had however reduced the amount of uninterrupted development hours. This was outweighed by the fact that the meetings created better visibility, oversight and knowledge, which allowed employees to tackle unforeseen challenges better, which had a positive effect on productivity as waste time was avoided [6]. Retrospectives are intended to support transparency and increase the productivity among others as a result of the project participants' learning from their own and others' mistakes, so that they are not repeated in the next sprint or iteration. Retrospectives should address the overall application of the method, its processes and practices, and the more specific experience in the daily development work and its relation to the resulting product [10]. The latter turned out to be an area the case unit did not focus on and thus did not benefit from [5, 6]. It can be explained with the case unit's early stage of utilizing Scrum and their lack of experience with regular retrospectives. Scrum of scrums should be used in large, complex development projects where several Scrum teams are associated [10]. The idea is to ensure transparency and the sharing and exploitation of potential knowledge that exists in different Scrum teams. The responses we received document that the way the case unit used Scrum of scrums had affected communication between the teams in a positive direction despite the fact that no clear guidelines for this communication activity had been developed [2]. Although it remained unclear whether the case unit used Scrum of scrums, as the literature recommended it, the use of Scrum of scrums did not only contribute to improve lines of communication in general, but also had a positive influence on the ability to monitor project progress. The interaction between the Scrum teams raised the common understanding of the individual Scrum team's status, where they positioned themselves as compared to the other teams and with regard to the development process of the overall product. Our overall positive assessment of Scrum on process transparency of the ISD process confirms empirically the expectations and claims, which are made in many of the conceptual and non-academic writings we had identified in our literature review. It also fills a gap in the area of empirical studies of agile software development [16]. In the absence of quantitative data and with no possibility to make direct measurements and collect such data throughout the project it is however built on subjective perceptions.

Nonetheless, on a more theoretical level our study can be related to complex adaptive systems (CAS) theory to find support for the increase of process transparency as one outcome of Scrum. CAS theory underpins agile ISD methods [17] such as Scrum and the case unit appears to be rather successful after its transition to Scrum. On this background the above results can be linked to CAS concepts and principles. If ISD, in our case agile development supported by Scrum, is understood as CAS, certain characteristics of the process are recognized to facilitate good performance, while others inhibit it [8, 18]. A number of concepts are frequently used when discussing CAS. They are intertwined and mutually reinforcing and have been put forward within the area of ISD as follows [9, 19]: Interconnected autonomous agents are able to independently determine what action to take, given their perception of their environment; yet, they collectively or individually are responsive to change around them, but not overwhelmed by the generated information flow. Self-organization is the capacity of these agents to evolve into an optimal organized form, which results from their interaction in a disciplined manner within locally defined and followed rules. Co-evolution relates to the fact that a complex adaptive system and/or its parts alter their structures and behaviors in response to their internal interactions and to the interaction with other CAS where adaptation by one system affects the other systems, which leads to reciprocal change where the systems evolve individually, but concertedly. Time pacing indicates that a complex adaptive system creates an internal rhythm that drives the momentum of change, which is triggered by the passage of time rather than the occurrence of events; this stops them from changing too often or too quickly. Poise at the edge of time conceptualizes a complex adaptive system's attribute of simultaneously being rooted in the present, yet being aware of the future and its balance of engaging exploitation of existing resources and capabilities to ensure current viability with engagement of enough exploration of new opportunities to ensure future viability. Poise at the edge of chaos describes the ability of a complex adaptive system to be at the same time stable and unstable; this is the place not only for experimentation and novelty to appear, but also for sufficient structures to avoid disintegration; CAS that are driven to the edge of chaos out-compete those that are not.

The above analysis has provided examples of interacting interconnected autonomous agents, such as the involved developers and testers, their self-organization as individuals and as project teams, their co-evolution through knowledge sharing and learning from each other, as well as for time pacing in the short iterative development cycles, and for poise at the edge of time and chaos, for instance with regard to compliance to deadlines and uninterrupted workflow, which thus empirically and theoretically lend support to the identified perceived positive impact of Scrum on process transparency and as indicated productivity, quality, employee satisfaction [6] as well as team leadership [7] of ISD in our case setting.

7 Conclusion

The normative literature on agile ISD [10] postulates the importance of process transparency for ISD projects and puts forward several ways such transparency should be achieved, but provides little empirical evidence how and with what effect transparency is accomplished in practice. We fill this gap and offer empirical evidence and theoretical backing of how process transparency is achieved and what its impact is. As such our work responds and contributes to calls for research on visual management [20] and the visual dimension of organizing [21] and supplements work on visualization related to IS on process flow charts and PowerPoint as representations and inscriptions in IT projects [22, 23]. It also extends work on the role of physical artifacts in agile software development [24] which hints both at the mutually intertwined and supportive, social and material, notational nature of story cards and wall boards. This work together with research on how development, implementation, and use of digital workflow visualization boards entangled in sociomaterial practices shape the focus of attention and facilitate integration of new operational practices [25] open up for further research on ISD in general and agile ISD in particular as sociomaterial practice [26].

While the usual disclaimers for the shortcomings of qualitative research also apply for our study, our work contributes to the body of knowledge in ISD with an empirical investigation that demonstrates the positive impact of Scrum on process transparency in ISD and project management and it provides a useful operationalization of the concept through five indicators. Despite the fact that the case unit had challenges, the indicators identified the areas where the company had achieved to exploit the potential of Scrum and its practices with regard to increasing process transparency and its effects. We also discussed how and with which impact process transparency was related to the other concepts of our framework. Although several authors underline the importance of an open organizational culture for agile development [9, 27] and argue that an open and innovative organizational culture is necessary to develop software according to agile principles we decided to disregard the concept as such as we assumed that the culture, its elements, the basic assumptions held by all members of that culture, their values and beliefs, and their artefacts and creations [28] and the cultural changes as a result of an implementation of Scrum would have an impact and become evident through the indicators. In other words, for culture as a broad concept we thought it would make more sense to be implicitly investigated through the process transparency indicators. In hindsight the relationship between culture and process transparency in the use of agile methods such as Scrum does however also merit a thorough investigation through future research on its own.

References

1. Johansen, T., Uldahl, A.: Measuring the impact of the implementation of the project mangement method Scrum (in Danish). M.Sc. Thesis, CBS, Copenhagen (2012)
2. Hirschheim, R., Klein, H., Newman, M.: Information systems development as social action: theoretical perspective and practice. OMEGA **19**(6), 587–608 (1991)
3. Newman, M., Robey, D.: A social process model of user-analyst relationships. MIS Q. **16**(2), 249–266 (1992)
4. Kautz, K., Madsen, S., Nørbjerg, J.: Persistent problems and practices in information systems development. Inf. Syst. J. **17**, 217–239 (2007)
5. Kautz, K., Johansen, T., Uldahl, A.: The perceived impact of the agile development and project management method scrum on information systems and software development productivity. Australas. J. Inf. Syst. **18**(3), 303–315 (2014)
6. Kautz, K., Johansen, T., Uldahl, A.: Creating business value through agile project management and information systems development: the perceived impact of Scrum. In: Bergvall-Kåreborn, B., Nielsen, P.A. (eds.) IFIP 8.6 WC on Creating Value for All Through IT, pp. 150–165. Springer, Berlin (2014)
7. Kautz, K., Johansen, T., Uldahl, A.: The perceived impact of the agile development and project management method scrum on team leadership in information systems development. In: Vogel, D., et al. (eds.) Transforming Healthcare through Information Systems, 24th ISD, pp. 167–183. Springer, Berlin (2015)
8. Kautz, K.: Beyond simple classifications: contemporary information systems development projects as complex adaptive systems. In: 33rd ICIS. Orlando (2012)
9. Highsmith, J.: Agile Software Development Ecosystems. Addison-Wesley, Boston (2002)
10. Schwaber, K.: Agile Project Management with Scrum. Microsoft Press, Redmond (2004)
11. Schwaber, K., Beedle, M.: Agile Software Development with Scrum. Prentice Hall, Upper Saddle River (2002)
12. Webster, J., Watson, R.T.: Analyzing the past to prepare for the future: writing a literature review. MIS Q. **26**(2), 13–23 (2002)
13. Moe, N., Dingsøyr, T.: Scrum and team effectiveness: theory and practice. In: Agile Processes in Software Engineering and Extreme Programming, pp. 11–20. Springer, Berlin (2008)
14. Sutherland, J., Altman, I.: Organizational transformation with Scrum: how a venture capital group gets twice as much done with half the work. In: 43rd HICCS. Kauai (2010)
15. Andersen, I.: The Apparent Reality (in Danish). Samfundslitteratur, Frederiksberg (2006)
16. Dybå, T., Dingsøyr, T.: Empirical studies of agile software development: a systematic review. Inf. Softw. Technol. **50**(9–10), 833–859 (2008)
17. Highsmith, J.: Adaptive Software Development: A Collaborative Approach to Managing Complex Systems. Dorset, New York (2000)
18. Meso, P., Jain, R.: Agile software development: adaptive systems principles and best practices. Inf. Syst. Manage. **23**(3), 19–30 (2006)
19. Vidgen, R., Wang, X.: Coevolving systems and the organization of agile software development. Inf. Syst. Res. **20**(3), 355–376 (2009)
20. Bell, E., Davison, J.: Visual management studies: empirical and theoretical approaches. Int. J. Manage. Rev. **15**, 167–184 (2013)
21. Meyer, R.E., Höllerer, M.A., Jancsary, D., Van Leeuwen, T.: The visual dimension in organizing, organization, and organization research: core ideas, current developments, and promising avenues. Acad. Manage. Ann. **7**, 487–553 (2013)
22. Locke, J., Lowe, A.: Process flowcharts: malleable visual mediators of ERP implementation. In: Puyou, F.R., et al. (eds.) Imagining Organizations: Performative Imagery in Business and Beyond, pp. 99–126. Taylor & Francis, Abingdon (2012)
23. Yakura, E.K.: Visualizing an information technology project: the role of powerpoint presentations over time. Inf. Organ. **23**, 258–276 (2013)

24. Sharp, H., Robinson, H., Petre, M.: The role of physical artefacts in agile software development: two complementary perspectives. Interact. Comput. **21**(1/2), 108–116 (2009)
25. Hultin, L., Mähring, M.: Visualizing institutional logics in sociomaterial practices. Inf. Organ. **24**(3), 129–155 (2014)
26. Cecez-Kecmanovic, D., Kautz, K., Abrahall, R.: Reframing success and failure of information systems: a performative perspective. MIS Q. **38**(2), 561–588 (2014)
27. Kautz, K., Pedersen, C. F., Monrad, O.: Cultures of agility—agile software development in practice. In: 20th Australasian Conference on Information Systems. Melbourne (2009)
28. Schein, E.: Organizational Culture and Leadership. Wiley, San Francisco (2004)

Printed in the United States
By Bookmasters